the state of the nations' birds

CHRIS MEAD

foreword by Tony Soper
illustrations by Kevin Baker

Whittet Books

DEDICATED TO
Max Nicholson & Robin Ruttledge:
continuing to show us the way, on both sides of the Irish
Sea, and
Ken Williamson (1914-1977), founder of the BTO
Population section

First published 2000
Text © 2000 by Chris Mead
Illustrations © 2000 by Kevin Baker
Reprinted 2000
Whittet Books Ltd, Hill Farm, Stonham Rd, Cotton, Stowmarket, Suffolk,
IP14 4RQ

British Library Cataloguing in Publication Data. A catalogue record for
this book is available from the British Library.
ISBN 1 873580 45 2

Printed and bound in the UK by Biddles

CONTENTS

FOREWORD

BY TONY SOPER

As I read an advance copy of this book I am rolling about at sea off the coast of Antarctica. Outside the porthole petrels and great albatrosses flash past in stately progression. From the open decks I can see penguins porpoising on their way to the breeding beaches. In a couple of weeks' time I shall be at home in Devon with a very different community of birds. Yet, as in the Antarctic, some of them are doing well and some are in decline Chris Mead has done us all a signal service in this splendid book by reviewing the status of our bird populations and gazing into his crystal ball. Seeing birds in their own surroundings and trying to know something of the ups and downs of their lives is what birdwatching is all about, and Chris is the ideal field-guide.

Neither of us could be described as 'spring chickens', and we have seen huge changes in the birds of Britain and Ireland and their habitats. In this book Chris describes what has happened, is happening, and has a stab at what is going to happen. When we started to look at birds the Peregrine was being severely affected by toxic chemicals - now they are two-a-penny. Collared Doves were rarities on the verge of colonisation. Nowadays we worry about the dearth of such once common and much-loved species as Song Thrush and Skylark. Yet if I were at home now I would be looking out of the kitchen window and expecting to see Cirl Buntings on the bird bath - who would have predicted this species as a success story? Happily this is a bird Chris is optimistic about, saved by a combination of detailed research with co-operation by farmers.

This book will become a practical tool of reference for all of us with an interest in our breeding birds. It documents the changes which have affected them and their habitats over the last century. Chris is the perfect man for the job. He has spent a happy lifetime working with the British Trust for Ornithology, Britain's premier amateur bird research organisation, but he has always been a near genius in communicating the findings of dry scientific research in a way that strikes home to everyone. He has trawled through countless reports, consulted with such bodies as the RSPB and BirdWatch Ireland and hordes of specialists to write a definitive snapshot of what is happening to our birds. We hope for more success stories in future editions. It is sad to read of our failures with the Red-backed Shrike, Lapwing and Black Grouse but wonderfully uplifting to enjoy the successes with Red Kite, Dartford Warbler and Little Egret. Watch this space!

PREFACE

'It seemed a good idea at the time' is a sentiment that most authors would agree with as they start writing their book. For me this is still the case: even some 18 months after having had the initial idea. It takes a serious look at our breeding birds, their history, current status and prospects, at the end of one millennium and the start of another. I have been around for almost 60 years but the book is the product of hundreds of years of observation, recording and reporting by huge numbers of birdwatchers. This book, and our unrivalled knowledge of our bird populations, is thanks to their enthusiasm - and I have picked out three of them, Max Nicholson, Major Robin Ruttledge and the late Ken Williamson, as supreme exemplars to whom the book is dedicated. There is also a very extensive network of national, regional and local organisations but, of course, the interpretation and conclusions are mine. I have tried to keep 'on side' with such organisations as the British Trust for Ornithology, Royal Society for the Protection of Birds, Irish Wildbird Conservancy and English Nature, but readers should not take the views expressed as anything but my own.

I hope that readers will be able to see how the history of our islands, and especially the influence of our species, has shaped the avifauna that remains. The detailed species treatments (and their boxes of statistics) are going to be what most people will consult regularly. These contain many sad stories, and some inspiring ones, but the reasons behind what has happened are explained in more detail in the habitat chapters. My personal postscript chooses the Red Kite as the bird of the century and the Red-backed Shrike as the biggest loser. After reading the book you may choose other species but I would expect you will agree with me - *we should have done better*!

There are many people who deserve special thanks. First of all Kevin Baker for the illustrations, which leaven the text, and Mike Read for the soaring Red Kite on the cover. Malcolm Ogilvie, as secretary to the Rare Breeding Birds Panel, was a great help and many other people let me have details of their findings in advance of publication. The BTO and T & AD Poyser allowed me to use statistics from their books and reports and the IWC provided the information for the *shamrocks* - due to be published just before this book. Much of the text was read, and commented upon, by Ken Smith and Steve Newton from the ornithological viewpoint. My brother, Stephen, wielded the red pen with a view to losing some of the most convoluted constructions and my wife managed to excise several dozen *howevers*! Annabel Whittet was not only my publisher but edited and set the text - delivered by e-mail - and supplied excellent free-range eggs. Finally Graham Evans, of Jacobi Jayne & Company, arranged for the book's appearance on the web - at www.birdcare.com/birdon

INVESTIGATING OUR

BREEDING BIRDS

There are literally millions of people who have an interest in birds in Britain and Ireland. The combined observations of a good number of them have enabled us to get a very good handle on what birds we have breeding, where they are doing it and how their populations vary from year to year. For a few species this is down to one or two real enthusiasts, for many more there are specialist study groups and for all there are the big and extensive projects. These are described below. All are on-going (but not annual in the case of the Breeding Bird Atlases) and all can do with more volunteers. If you would like to participate yourself the description of each project contains – at the end – the information you will need.

The National Breeding Bird Atlas

In the mid-1960s the idea of producing a *Breeding Bird Atlas*, based on the 10-km squares of the National Grid, was proposed and took shape. The BTO(British Trust for Ornithology) decided against any idea of trying to assess the numbers of birds in each square and a presence or absence system was agreed. Three categories of record were decided upon – proved breeding, probable breeding and present in a possible breeding area. The *Atlas* ran from 1968-1972 and achieved excellent coverage and was published in 1976, edited by Tim Sharrock as a joint publication of the BTO with the IWC (Irish WIldbird Conservancy). The *New Breeding Bird Atlas*, covering the years 1988-1991, came 20 years later and included much more information. Each species has the new map of occurrences and a map showing the changes that these represent from the first Atlas. In addition many species have abundance maps showing the relative abundance across Britain and Ireland computed from timed records in several tetrads (2x2 km squares) within each 10-km.

Huge numbers of volunteers spent almost 100,000 hours on the timed visits alone – very nearly an average of 24 hours for each of the 3,858 10-km squares covered. The combined timed and untimed records came to more than 550,000! The coverage in areas away from centres of population was probably not quite as good in the second *Atlas* as in the first and this should be borne in mind when looking at the comparisons given in the boxes printed in this book. The *New Atlas of Breeding Birds in Britain and Ireland: 1988 –1991* was published in 1993, edited by David Wingfield Gibbons, James Reid and Robert Chapman who had been the organisers for the BTO, SOC and IWC respectively.

In this book the top line of each box refers to the *Atlas* data. The number of 10-km squares in Britain and in Ireland are given with, in brackets, the percentage change since the first *Atlas*. These changes are probably not so reliable for crepuscular and nocturnal species which were not as well recorded in the second *Atlas* in some areas. The recorders were asked to concentrate on a daytime timed record rather than try for evening or night-time visits. There are dozens of local breeding bird atlases both published and in progress. These are invaluable for local bird watchers and fascinating to do – I have not tried to scan them as this book is concerned with the national picture. To find out what is going on in a particular area contact the BTO Regional Representative for the area. There is a great deal of national publicity when a national atlas is being undertaken – and these may, in the future, not just be in the breeding season. We have already had a winter one.

Rare Breeding Birds Panel

Started in 1972, the Rare Breeding Birds Panel (RBBP) tries to record the numbers and localities of rare birds breeding in Britain and Northern Ireland. It has gradually earned the respect of bird watchers and now receives information on the vast majority of species. Two exceptions are Honey Buzzard and Goshawk where there are several observers who are not yet co-operating. Records are treated in the strictest confidence and an annual report appears in the monthly journal *British Birds*. The latest data available covers 1997 and, for the first time, records of exotic breeding birds were also collected for 1996 and 1997. Exotic birds are species not admitted to the British List as having established feral populations but they are breeding, apparently wild, as a result of being introduced, deliberately or by mistake, by man.

The panel has seven members who serve in a personal capacity but four of them also represent the interests of the four bodies who sponsor the RBBP: the Joint Nature Conservation Committee (JNCC on behalf of the country conservation bodies), the RSPB, the BTO and *British Birds*. Most records are forwarded to the panel through the county recorder system – with which all serious bird watchers should co-operate. In each report estimates are made of the breeding numbers with a minimum and maximum value.

The existence of the RBBP is a very important part of the protection of the rarest species. All the information should be in one place and curated by a small group of experts with impeccable credentials and the trust of the birding community. Species that the RBBP is interested in are changed at intervals – as birds become rarer or more common – and are regularly listed in *British Birds*. Any new breeding species, or bird which is present in the breeding season in suitable breeding habitat, will also automatically be included. There is not yet an equivalent of the RBBP in Ireland.

RBBP data have been used for the numbers breeding in Britain for most of the species reported to the panel. They have also been used in calculating the British population trend. The numbers given in this book from species study groups and other up-to-date sources have the year to which they refer in brackets in the text. The Panel can be contacted through *British Birds*, your local bird recorder or the current secretary is Malcolm Ogilvie, Glencairn, Bruichladdich, Isle of Islay PA49 7UN.

The EBCC Atlas of European breeding birds

Arguably the biggest scientific project on wild birds ever undertaken, this project aimed to map all Europe's breeding birds – from the Mediterranean coast in the South to Svarlbad and Novaya Zemlaya in the North and from the Azores in the West to the Urals in the East. Squares used for county bird atlases are small with 2-km sides, the National atlas has 10-km side (100 square kilometres) and the ones for the EBCC are 50 km x 50 km – 2,500 square kilometres. This is a bit more than the size of Hertfordshire! The results were published in 1997 and included population estimates for each species for each country. Information for the *Atlas* was gathered through regional organisers and not through individuals.

The European Status information is their estimate of the population for Europe (this is a geometric mean and is printed to two significant figures). The percentage in Britain and Ireland is just that – using the figures for Britain and Ireland from the EBCC database. Any number (1-10) appearing after this is the ranked figure giving the importance of Britain and Ireland's population in relation to all other European countries: =4 means that three countries have a higher estimated population and one is the same as ours. For this pan-European survey you take part through the National Atlas surveys as and when they take place rather than as an individual.

Common Birds Census & Waterways Bird Survey

These are surveys organised by the British Trust for Ornithology to provide a population index for breeding birds in farmland, woodland and beside water. The idea of a population index is to try to find out how the total number of birds varies from year to year without knowing how many birds you are dealing with! It is obviously impossible to go out and count every single pair of Skylarks every year – there are over a million breeding in Britain even after many years of decline. However by carefully mapping the territories that are occupied on one or two hundred plots, scattered through the countryside, and then doing it again on the same plots, with the same rules and the same people recording, one can get a very good idea of the change from year to year.

Over the years the Common Birds Census (CBC) has built up a formidable body of information and really good population change information for a number of species. Analytical methods have been developed; now the information for each year is carefully considered with the data from up to six years to smooth out violent changes.

It was realised that there were difficulties. For instance the only two habitats covered well were farmland and woodland. The plots chosen were selected by the CBC workers themselves and they often built up a special relationship with the land's owners. This could result in the figures from their plots being unrepresentative of the country as a whole. There was also the suspicion that plots were likely to be dropped if they were so severely changed – for instance by intensification of agriculture – that they were no longer fun to cover. It was also clearly rather difficult to follow the fate of species that were not territorial but colonial.

In 1974 one of these problems was addressed by the start of the CBC's sister scheme, the Waterways Bird Survey (WBS). This sought to extend the species covered by the CBC to include some of the water birds and those that congregated along the water's edge – riparian species. This scheme proved to work well along the edges of rivers, streams, canals, etc. Additional species were covered and alternate population indices were obtained for others. Again the plots are chosen by the surveyors so they are not necessarily representative of the country's water birds as a whole. For both surveys there was also a fundamental problem. Where there are lots of BTO members it is relatively easy to persuade some of them to undertake the work, but in the Highlands and Islands and other places with sparse populations surveyors are difficult to recruit and retain.

The boxed information on British population trend is based on the

CBC or WBS data over the 25 years 1972-1996 for CBC or the 23 years 1974-1996 for the WBS (WBS appears after the figure if it is based on WBS). The CBC data is based on all sites unless an **F** (Farmland) or **W** (Woodland) appears after it. The year 2000 will be the last when extensive CBC and WBS plots will be used for annual monitoring purposes and all efforts will be concentrated on the Breeding Bird Survey – see later. The very latest published changes (1970-1998) are in {curly brackets}.

Constant Effort Sites

The Constant Effort Sites scheme (CES) is only a small part of bird ringing. The system of putting small numbered and addressed metal rings on birds' legs is a hundred years old. It provides enormously important information of all sorts on all aspects of our birds' lives. For the breeding birds the detailed information on survival rates, from general ringing, is very important but the CES, started by amateur ringers almost 20 years ago, is very special. CES ringers are a breed apart. They ring at their plots every ten days throughout May, June, July and August. By recording the full details of age and sex and, of course, of the birds they retrap (catch wearing rings already) they are able to get to grips with the population levels of the adult birds, the productivity of the birds by working out the ratio of young birds and, finally, the survival of the birds from year to year. The CES started with 46 sites in 1983 and had over 130 by 1998. Any data which derive from the CES have those initials beside it.

To participate in the CES you must be a ringer – or be helping a ringer. There are ringers all over the country and many of them are qualified to train new ringers. The whole scheme is run by the BTO and is very carefully regulated to ensure the safety of the birds and the integrity of the data. Ringing carries a serious health warning as many of the people who start doing it have their lives taken over by their hobby. It is also not cheap as the scheme's costs are only supported in part by the Government contract, through JNCC, with the BTO. Ringers have to pay for the rings and equipment they use – although there are special subsidies for schemes like the CES which are deemed to be particularly important for conservation. Write to the Ringing Office, BTO, The Nunnery, Thetford, Norfolk IP24 2PU.

Nest Record Scheme

This is a really simple idea which has proved very significant. The BTO started it 60 years ago and it was designed by Julian Huxley and James Fisher. The concept was simple. A single person devoted to nest finding and recording could come up with a very good idea of the average clutch size and start date of several of their local species in a lifetime of recording.

But if there were hundreds of recorders all noting down the contents of all the nests they found in a standard way, and sending them to a central office, simple statistics could be calculated each year and for different regions.

There are now nest recorders all over the country and over a million cards in the files. The cards, if piled one on the other, would reach higher than the Post Office Tower! About 30,000 are received every year and there are more than 100 for 65 to 70 species. Many of them are computerised and all sorts of analyses are undertaken including annual measures of laying date, clutch size, brood size and daily failure rates for the egg stage, for fledging period and throughout the period from egg laying to fledging. These calculations are restricted to species with a decent sample of cards contributing to the analysis – at least 100. Recent analyses have proved the effect of Global Warming bringing forward the laying dates of many of our birds (see Climate and Weather chapter). Data deriving from the NRS have these initials beside it.

New nest recorders are always needed. Their results may be valuable even if they relate to only a dozen nests of common species well recorded in a garden or even six nests of a rarer species that the recorder is studying for other reasons. However some recorders regularly submit hundreds of records a year and there are several groups of friends who are continuing to make a big contribution year after year. In 1998 Government support for the NRS was cut as part of the overall contract negotiated with the JNCC. This does not mean that the NRS is any less important and so a donation of the cost of five Nest Record Cards is being made to the BTO, to help the scheme, for every copy of this book sold. To take part yourself write to Nest Records, BTO, The Nunnery, Thetford, Norfolk IP24 2PU for a starter pack.

The Breeding Bird Survey

In 1994 the newest all-species monitoring survey was launched on the geographical and species spread of breeding bird monitoring. The Breeding Bird Survey (BBS) is a partnership between the BTO, JNCC and RSPB and is much simpler to undertake than the detailed mapping of the CBC or WBS. The 1-km squares are selected using stratified random sampling techniques from 83 regions. Volunteers from the regions are allocated the squares they are to cover rather than choosing where to go. They need to make a fairly detailed habitat record, according to a simple but very comprehensive coding system. The bird recording takes place over about 90 minutes of an early morning in the first and second halves of the breeding season. Walking two 1,000 metre transects through the square every bird encountered is noted – together with the distance category from the transect line (within 25 metres, 25-100 metres, further or flying). The observers are also asked to record any mammals they encounter.

The latest report (1998) reported on 2,297 squares which were covered by the volunteers. These represented 64% of the squares allocated and were scattered throughout England (1708), Scotland (302), Wales (192), N. Ireland (82), Channel Islands (7) and Isle of Man (6). The total area covered was bigger than Hertfordshire, quite a large county! The BBS results only cover a short period, five years, but they are already showing significant changes. Where these are included the initials BBS appear. It promises to be an excellent monitoring scheme for widely distributed species but clearly, as was expected, it will not be good at covering colonial seabirds, for example.

New observers are needed as the percentage of issued squares covered is still only about 65% and new squares are being added each year. The Countryside Bird Survey(CBS), run by BirdWatch Ireland and the National Parks and Wildlife service was launched in 1998. It follows the methodology of the BBS and it was hoped that up to 300 squares would be covered in the Republic during the first summer. To help there contact BirdWatch Ireland, Ruttledge House, 8 Longford Place, Monkstown, Co Dublin, Ireland. For the BBS contact your local BTO Regional Representative or BBS, BTO, The Nunnery, Thetford, Norfolk IP24 2PU.

In the boxed information on Conspicuousness in summer the number of BBS squares covered in 1997 and 1998 where the species was recorded is given and the percentage (out of the total of 4491) and the rank that the species holds of all species recorded. After this ranking, which refers to the BBS, is the ranking of the species for the first two years (1998 and 1999) of the CBS in the Republic of Ireland [in square brackets].

Garden bird recording

Garden bird recording does not form a major part of this book but gardens do form an increasingly important habitat for our birds. Records from the BTO's long running Garden Bird Feeding Survey (GBFS) are included in Human Sites chapter. This survey completed its 29th winter of recording in spring 1999 and is restricted to 250 participants who send in their weekly records in considerable detail. There is a long waiting list for vacancies which occur. A simpler, year-round survey – Garden Bird Watch (GBW) – is now into its fifth year. This is self-financing and costs £10 p.a. for the recording forms and four copies of its own magazine *The Bird Table*. New recruits get a detailed summary of the first three years of observations covering a staggering 500,000 or more weeks of recording. Write to Garden Bird Watch, The BTO, The Nunnery,

Thetford, Norfolk IP24 2PU. In the Republic of Ireland the Winter Garden Bird Survey had covered 1,200 different gardens in its first five years. Forms are available from BirdWatch Ireland, Ruttledge House, 8 Longford Place, Monkstown, Co. Dublin and the recording period is from December through February each year.

BTO alert system

Many of the BTO investigations described in this chapter form the basis for BTO research which will be reported to the JNCC every three years in a weighty report. At the moment it is called *Breeding Birds in the Wider Countryside: their conservation status*. In it three types of change in species' parameters are highlighted. These are basically recommendations from the BTO and there may be other information, available from other sources, which will lead to the exclusion of a High Alert species from the Red List of Species of Conservation Importance. For instance the High Alert might be based on population trends from farmland but refer to a species whose major population inhabits upland areas. The designations in the latest report are as follows:

HIGH ALERT: Decline in population index of 50% or more over the last 25 years (CBC) or since the start of WBS (23 years) or CES (14 years) choosing the most appropriate scheme for the species. Or, for species where the population is not monitored accurately, a decline in breeding performance or survival which is likely to have a severe affect.

MEDIUM ALERT: Decline in population index of 25% (see above) or a statistically significant decline in breeding performance or survival.

RECOVERY: This is only used for species already on the Conservation Importance Lists due to declines measured by the CBC which would not, using the most up-to-date information, warrant inclusion. This does not mean, necessarily, that their former populations have been regained but that the steep decline has been halted.

MONITORED: some aspects of the population or other statistics are monitored by the BTO and the results do not warrant listing on either of the alert lists. If better statistics were available some species might be on the alert lists.

Specialist groups

There are all sorts of special groups interested in particular species or groups of species. These range from the Wildfowl and Wetlands Trust and the Game Conservancy – big membership-oriented organisations – through more specialist groups like the Wader Study Group and the Seabird Group. Others are concentrated on particular species – like the Golden Oriole Study Group – or a group of species in a particular area – like the Raptor Study Groups. Often these small groups have a particular person concerned with the conservation of the species that they are all interested in who works professionally on the species. In some cases the group of people does not 'exist' as a formal entity but all concerned co-operate and might produce a species newsletter at the end of each season. Their contact with the RBBP will usually be channelled through the group – so making Malcolm Ogilvie's life much easier as secretary to the RBBP. One particularly important, regular survey which takes place at 15 year intervals is in progress at the moment. Seabird 2000 is the third time that our seabirds have been censused in a huge and complicated operation mainly organised by the Seabird Group and JNCC Seabirds team.

The groups I have been writing about, up to this point, are desperately interested in the conservation and welfare of the birds. They study them to be able to further our knowledge and understanding, and hope and expect their observations to make a difference to the bird's future. Unfortunately this is not always the case. For instance the Jourdain Society is dedicated to the scientific study of birds through their eggs – oology or egg collecting. Whilst the aims of the society are not illegal several of its members have been prosecuted for offences against the Wildlife and Countryside Act. You will not find its address at the end of the book.

Likewise there are a few obsessive twitchers who feel that they have the right to see birds regardless of the consequences and try to discover where they can find breeding rarities simply to add them to their lists. In many cases, on reserves, this can be arranged conveniently for the visitors and with no inconvenience for the birds. However with birds on private property, with species liable to desert if disturbed or with species likely to be the target of egg collectors, falconers or bent gamekeepers, it is the height of stupidity to put them at risk. For example it is perfectly possible to see the introduced Red Kites on the Chiltern ridge from pub beer gardens around and about Watlington and Stockenchurch or from the Aston Rowant Reserve. There is no need to clog up the tiny local lanes or annoy local people by trespassing in the woods where they nest. Indeed that would be illegal under the Wildlife and Countryside Act.

As an undergraduate at Cambridge 40 years ago I was in a car driving past the Ouse Washes in May. The four keen birders on board convinced each other that there was no point in stopping and looking. Several years later, when the news that Black-tailed Godwits were breeding was released, we discovered that we had all known about this. However we did not know that the others knew and none of us ever stopped to look in case it drew attention to the site!

And finally ...

The summaries in this book have been compiled through the efforts of tens of thousands of bird watchers. It tries to be as complete as possible but does not include records which are being suppressed by the conservation community. This is not the publication in which to find the details of the nesting of Great Auks under Wigan Pier!

Brown, A.F., Stillman, R.A. & Gibbons, D.W. 1995 'Use of breeding bird atlas data to identify important bird areas: a northern England case study' *Bird Study*: **42**, 132-143.

Donald, P.F. & Fuller, R.J. *1998* 'Ornithological atlas data: a review of uses and limitations' *Bird Study*: **45**, 129-145.

Gilbert. G., Gibbons, D.W. & Evans, J. 1998 *Bird Monitoring Methods*. RSPB (with BTO/WWT/JNCC/ITE/Seabird Group).

Hickling, R. (for the BTO) 1983 *Enjoying Ornithology*. T & AD Poyser.

Peach, W.J., Baillie, S.R. & Balmer, D.E. 1998 'Long-term changes in the abundance of passerines in Britain and Ireland measured by constant effort mist-netting' *Bird Study*: **45**, 257-275.

Siriwardena, G.M., Baillie, S.R. & Wilson, J.D. 1998 'Variation in survival rates of some British Passerines with respect to their population trends on farmland' *Bird Study*: **45**, 276-292.

Stone, B.H., Sears, J., Cranswick, P.A., Gregory, R.D., Gibbons, D.W., Rehfisch, M.M., Aebischer, N.J. & Reid, J.B. 1997 'Population estimates of birds in Britain and the United Kingdom' *British Birds*: **90**, 1-21.

Mute swan.

FARMLAND

For most people in Britain and Ireland the countryside that they are most likely to see and visit will be farmland. This chapter looks at lowland farmland – both arable and pasture – as one of the most altered habitats within Britain over the last century. Many of its characteristic birds have suffered major declines over the last 30 or 40 years and the changes for the human population have been drastic too. Upland habitats are often just as much farmland as lowland areas but these are dealt with in the Uplands Chapter.

Farmland is an entirely man-made habitat which has evolved over the last few thousand years – but never so fast as recently. It has replaced all sorts of natural habitat usually involving the destruction of woodlands and wetlands. Most of the rolling fields we see now were once covered in woodland and many of the flat areas were wetlands – either carr where trees predominated, reed-beds, open water or saltmarsh. These natural habitats each had their own rich assembly of breeding birds that had gradually evolved over many thousands of generations. The birds that found a home on farmland had aspects of their lives which suited them to the new conditions. Many of these were birds of the woodland edge which were able to utilise the hedges and copses, others were birds of open country that colonised the open fields. In Ireland pasture has always predominated and the potato famine, 150 years ago, was the defining moment in recent history, as death and emigration drastically reduced the human population, and much land went out of agricultural production.

However farmland, as we think we know it this century over much of Southern Britain, is very different from what we would have found 300 or 400 years ago. Before the enclosures boundary hedges were rather rare. They were generally only there to mark important boundaries between parishes, around parkland and along drove ways. Cultivated land was tilled by hand, with horses or with cattle drawing the plough. Cattle, sheep and pigs were grazed under the supervision of men and boys and so did not need hedges to keep them in one place. Very often common land, sometimes woodland, was used for pasture. The number

of people employed on the land was huge and it looked – and was – very different from now. Land holdings were large but the rural economy depended, to a very large extent, on the human resource so that the large land owners had to keep their villages going and were not able to run their farming enterprises by buying bigger and better machinery!

In the Middle Ages the major event was the Black Death, starting in 1348, which reduced the numbers of people available to till the land and led directly to the rise of the sheep. There ensued a certain amount of enclosure and a tremendous increase in wealth in some areas – visible still through large and magnificent churches in many areas dating back to this time. In the North and West the open field system, without hedges, was not normal. In the South and East the open field system, without hedges, was prevalent, which would have favoured open country species like Grey Partridge, Lapwing and Skylark. This was replaced by the main Parliamentary Enclosures, allowing for the enclosing of common land leading to small fields with hedges as boundaries. The Enclosures took place mainly from 1760 to 1815. Most of the areas enclosed were already arable but about a third were not – about 3 million hectares were involved.

At about the same time the arable cultivation system was transformed by the idea of the Norfolk and, later, other rotations. Instead of a third of the land being left fallow and without a crop each year to recover its fertility, a four-yearly succession of different types of crop enabled the farmer to obtain useful production each year. The classic rotation was clover for fodder and fertility; wheat; roots for cleaning the crop; and barley. This also shifted the focus of arable farming from heavier clay soils to lighter soils where the rotation worked well. The changes to the avifauna over this period would have been immense particularly as the system of hedges criss-crossing the countryside developed and matured. In particular birds from the woodland edge would have been able to exploit a much wider area of the arable land as the hedges spread.

At the start of this century horse power still reigned supreme although there were steam ploughing and threshing machines in operation and, in some places, oxen were still used. Very large numbers of people were still involved with farming although the seeds of the farming slump of the 1920s and 1930s had been sown 20 years earlier with free trade allowing cheap, high quality food to capture what were purely domestic markets. During the slump many fields were not planted with crops, and grassland – of various types – comprised over 80% of farmland. The changes in use of the areas of farmland, stocking rates and yields are listed in the tables which are from O'Connor and Shrubb (1986).

Since the early 1980s there have been huge changes with millions of cattle being slaughtered through the disaster of BSE and the rise and fall of set-aside. Whilst the changes over the last few thousand years have been gradual, the changes this century driven by the economy, by politics and by technology have often been very swift. They have altered the population levels and mix of species of birds over much of Britain and Ireland.

The agricultural slump

The economic forces that affected agriculture up to a few years before the start of the Second World War tore the heart out of the industry. The Depression, affecting the Stock Market, was not the cause; rather, the ability, through fiscal change and modern transport, of food manufacturers to import cheap food from abroad caused most producers severe difficulties. This happened long before the Depression. Arable farmers were unable to produce quality wheat at prices which could compete with the grain provided by the huge fields in Canada and the United States. Dairy farmers were unable to produce milk for processing as butter and cheese to match the cheap, and good quality, product of New Zealand. And livestock farmers were in direct competition with excellent cheap lamb from New Zealand and beef from Argentina. Things became slightly better for a few years at the end of the First World War when grain prices were guaranteed for a short time. The Depression compounded their problems as fewer and fewer people were able to afford to pay premium prices for home grown produce.

Breakdown of agricultural land

(Areas given in hundreds of thousands of hectares)

ENGLAND	Pasture	Rough grass	Arable
Late 1800s	41.5	11.6	54.7
Early 1930s	54.9	18.2	35.6
Early 1960s	39.0	13.1	52.2
Early 1980s	31.8	10.1	52.7
SCOTLAND			
Late 1800s	4.4	43.1	14.0
Early 1930s	6.3	44.7	12.3
Early 1960s	3.5	50.2	13.8
Early 1980s	5.6	42.6	11.2
WALES			
Late 1800s	7.1	5.7	4.5
Early 1930s	8.5	8.7	2.5
Early 1960s	7.0	6.6	3.4
Early 1980s	8.4	5.3	2.6

ARABLE	Cereals	Ley grass	Roots
Late 1800s	38.2	17.4	15.0
Early 1930s	20.5	15.9	10.5
Early 1960s	30.3	25.9	10.4
Early 1980s	39.2	16.3	7.4

STOCK & YIELD	Cattle 100,000	Sheep 100,000	Wheat Tons/ha
Late 1800s	60	296	2.02
Early 1930s	73	253	2.02
Early 1960s	107	281	3.87
Early 1980s	117	317	6.05

The effect on the industry was catastrophic. The table shows that the yield of wheat did not change over 50 or more years. There were few 'artificial' fertilisers, no clever chemicals to control pests, no new varieties of crop to allow regular autumn sowing, few tractors and no combine harvesters in general use. These 50 years did not see major developments to alter arable or livestock husbandry. Things just got harder and harder for the farmers and, especially, the farm workers. What was particularly difficult was that so many of the labourers lived in tied cottages so that when they lost their jobs, as many did, they lost their homes as well. The lack of a market meant that much arable land was lost to cultivation and became pasture or rough grass (see table). Yields of wheat remained the same over these years. The stocking rate also remained similar with the national cattle herd increasing by 22% and the sheep flock decreasing by 16% as the amount of pasture went up by 16%. There may have been effects on the birds but they are not likely to have been very important for many of them. The fields remained the same size, their hedges may have been cut less often and fewer were cultivated. The situation was about to start to change rapidly.

The rise and rise of arable: World War II

The farming industry began to recover beforehand, but the war, and the great need of the nation for home-grown food, provided it with an enormous boost. The use of shipping to transport food from abroad had to be weighed against the use of the ships for oil, other raw materials, weapons and munitions. Farmers were encouraged and required to bring back their holdings from grass to arable and to plough up areas – like downland and meadows – to plant crops. This was supervised by the War Agricultural Committees. They knew their local areas well and put pressure on the landowners to do their bit. Some farmers already had tractors and were using bigger ploughs, cultivators, reapers and other machinery. Wet areas were drained for cultivation and production and productivity increased. Some bird species will have suffered where downland was ploughed and became arable fields rather than sheep walk and where marshes were drained and taken into cultivation. However the fields themselves remained very much as they had been. The weeds and insects within the crops provided food for the birds and many areas were left as winter stubble and fallow land in the summer. The massive effects on the birds were about to start.

Bringing in the chemists and technicians

Chemical fertilisers were widely introduced in the 1930s but these were not the substances which were to alter the face of agriculture completely. As a direct result of the war new insecticides were introduced which provided the farming industry with the first in their present-day armoury of chemical tools. The technicians were doing their bit as well. Big changes were being made with the introduction of new, much bigger, machinery and newly developed varieties of crop and animal. Fifty years ago saw

Sparrowhawk and Robin.

the beginning of the rapid changes in agriculture and their effects on birds.

The first complex chemicals, with very long-term effects, were the organo-chlorine compounds. These included DDT, Gamma-BHC, Aldrin and Dieldrin amongst others. They seemed – and were – life savers when they were introduced to kill the insects which transmitted diseases. Their use in agriculture to control (kill) pests was an inevitable development. Their persistence was also looked upon as a really positive factor. It meant that the time-consuming and costly application process was not going to be needed so often. But the persistence proved to be an horrendous problem as the pesticides not only killed directly but also built up in the food chain.

Birds like Woodpigeons, Stock Doves and Pheasants died from, for instance, eating dressed seed. Those birds that had eaten the affected seed began to ail and were thus more likely to be selected by predators. The residue was then inevitably passed on to their predators. So, suddenly, the populations of Sparrowhawks and Peregrines crashed as the chemicals ingested in their food built up in their bodies. These birds also suffered as the chemicals interfered with their calcium metabolism and their eggshells became so thin that the eggs broke: even the weight of the incubating female caused them to collapse. Many of the adults seemed perfectly healthy but they were unable to breed. The birds disappeared from areas near and on farmland, and the populations of upland birds were severely affected by the same chemicals being used in sheep dips. The chemicals were proved to be doing the harm after about ten years and were banned. Peregrines recovered after about 25 years and Sparrowhawks are back almost everywhere 30 years later.

But the effect of farm chemicals has caused so much more damage to the bird populations than the obvious and well publicised effects of the persistent organo-chlorine family in the 1960s. There have been (and still are) direct mortalities from chemicals old and new but the really staggering effect is on the ecology of farmland. A part of the primary productivity, the transfer of energy from the sun into life forms, always used to be utilised outside the crop. Weeds grew in the fields, fungi grew on the crop, insects used the weeds and the crop plants; the birds ate the insects, and the weeds and their seeds in the winter. Now that has all changed. *All* the productivity in a field is directed to the crop. The fungi are controlled by fungicides, the weeds are controlled by herbicides and the insect pests by pesticides. It is little wonder that our farmland birds are suffering.

In 1997 I walked 600 metres down a tramline – where the wheels of the tractor go through the crop – in a field of wheat. The only plants I encountered were wheat except for four individual plants of goose grass. There were no visible insects and the birds were all around the field margin. The cropping rate for that wheat would now be in excess of the 6 tonnes per hectare average logged in the early 1980s – a three-fold increase in 50 years. Were just 2% of that productivity, from just one hectare, to go for wildlife it would probably be able to sustain more than 10 families of Skylarks through the year! The maths are simple: 120 kg of wheat would provide 20,000 daily feeds of 6gm. A pair of Skylarks might raise 8 young but these 10 birds would gradually dwindle until, by the next breeding season, only two would mature. On average, a pair and its progeny would need about 2,000 daily feeds. So 100 pairs of Skylarks could survive on a single 10 hectare field which would still produce over 60 tonnes of grain! Of course all the productivity going to wildlife would not be directed at the Skylarks but the birds are able to exploit both plants and insects.

Soil invertebrates are also targeted by the chemicals and many suffer from compounds used for other purposes. Birds like Starlings, thrushes and Lapwings are particularly affected by these losses all through the year. The lack of insects in the crop and in the hedges round the crop, so often just as bare of life because of spray drift, is the reason that Grey Partridges have done so badly and has further effects. Many of the birds of farmland, whether the adults eat seeds or insects, rely on small, soft-bodied invertebrates whilst they are feeding their young.

These changes spread through other crops, but the use of herbicides was delayed in root crops because the earliest formulations were toxic

Black-headed Gulls.

to the crop plants. Technical developments have solved these problems and now the rich assortment of weeds that used to grow amongst the sugar beet, mangolds and turnips is a feature of the distant past – and the large flocks of finches that used the resource have no crops where they can go.

In winter open ploughed land also provided these seed-eating birds with a long-term food resource as there was a replenished seed-bank in the soil from the weeds; efficient herbicides have meant that this resource is now not replenished each year. Some weeds have seeds which can remain viable for many years – see how a miss in herbicide can cause a gash of red poppies across a field that has had none for 10 or 20 years. However the mass of useful seeds within the soil as food for the birds dwindles much more quickly. The best estimate is that the seed bank in arable land has dropped to less than a tenth of what it was 50 or 60 years ago – from 20,000 per square metre to 2,000 or less.

Men and machines

Farming has been man and machine trying to tame nature ever since the first digging sticks were used. The plough and the use of animal power tipped the balance away from nature, but it was not until powered machinery started to be used as a matter of course that real changes began to make themselves rapidly apparent. The tractor for ploughing, the combine for harvesting and the sprayer for applying chemicals have enabled the work on the land to be done by fewer and fewer people. The percentage of the total population in England and Wales occupied in agriculture was 20% in 1860 and had dropped to 3% by 1950 – and is still dropping.

There were further consequences of the use of big machinery: to operate at its best it needs a decent run at its tasks. There is no point in having big combines and then spending hours each day dismantling them to get the machinery into many little fields. Whip out the hedges – and gain a bit of land – and run a very efficient pattern of cultivation throughout the year. Most of the farmland birds are related to the hedges and not to the open field and the loss inevitably led to many birds declining as the available habitat has declined.

But, of course, this is not all. Hedges with trees could shed a branch into the crop and a combine driver, not keeping a proper lookout, might damage the machinery on it. This is not only an extra, and avoidable, expense but also deprives the farm of the use of a very expensive machine for hours (even days) for the short period that it is needed each year. Even the production of vegetables, like carrots, has been developed out of all recognition: huge machines make sure that the beds the roots are grown in are stone free. Many of them are also grown with the use of miles and miles of plastic mulch. The modern spraying machinery covers a bigger and bigger swathe of the crop and often spray drift affects the hedges. Plant diversity becomes impoverished and insect diversity drops like a stone. This means that there is less food for the birds: a further factor causing declines.

Crops and cropping

Many new crops have been introduced in recent years. Some are obvious like the springtime yellowing of the countryside as oil-seed rape strides across it or, on light soils, the summer morning bright blue haze of the flowering linseed. Ripening and spilt seed from both these crops are excellent food for small birds but are normally only available for a short time each year. One consequence of the change to oil-seed rape has been the colonisation of some of the fields by nesting species like Reed Bunting and Sedge Warbler. These are often successful provided the ripe rape is desiccated and not mown before harvest. Maize, grown as a forage crop and also for cobs, and even sunflowers and lupins are grown as field crops. However just as different are the modern varieties of wheat and barley. They grow on shorter, stronger straw, whose growth may be chemically regulated, and are bred not to shed their grain. Many can be grown from autumn and not spring sowing. The hay meadows have been replaced by fields growing rye-grass for silage, which are cut earlier and earlier each year. This means that the species nesting in them do not have long enough to complete their breeding cycle. Even where the crop is still hay it is only in the remotest areas of the West in Scotland and Ireland that the Corncrake can still breed – and even then only with special measures taken for its well being. A hundred years ago the birds were common throughout the country.

An immediate consequence simply of the difference in timing of sowing is that there are millions of hectares less stubble and open ground available for the seed-eating birds to live on in the autumn and winter. The fields have been ploughed and planted in August or September and are green with the new crop over the winter. Indeed that crop becomes much denser than the old spring sown crops and is more difficult for the crop-dwelling birds to live in during the spring. Lapwings are unlikely to nest successfully and autumn-sown fields are shunned by breeding Skylarks who have any sparser alternatives in which to nest. However the problem in autumn and winter is compounded by the depletion of the food resource within the existing stubble. The only places left are where spring sown crops are to be drilled or the field is left as set-aside.

In times before combines picked up every last ear, even if the crop was laid, lots of the grain grown by the farmer was not harvested. Over-ripe heads in the field shed their contents and it was very much worth while for the local people to glean the stubbles for waste grain for their chickens. The spilt grain was a very useful resource for the wild birds, such as Skylarks. Individual grains of wheat, for instance, are so big that a Skylark would only have to find 100 each day to stay alive. But, even without the spilt grain, there used to be lots of weed seeds available for the birds in the stubbles; these are down by over 90% on average. The search time needed to find food on many stubbles is probably now uneconomic and the unfortunate Skylark would need too long to find the food – in relation to the shortened daylight hours of winter – to be able to survive.

These, and other effects, are most notable in southern and eastern Britain but their consequences are also felt elsewhere. More and more areas are able to grow crops that used to be restricted to southern areas. New varieties have been introduced, produced by selective breeding, that are able to tolerate shorter growing periods and lower temperatures. Oil-seed rape, for instance, could only be grown in southern Britain but is now a common crop as far north as Inverness – even on Orkney – and west to Wales and

Ireland. Unnoticed by many people, the grass being grown has also been developed so that it can be cut earlier and more often. This has not affected the far West where the late growing season and poor soils cannot support the modern varieties. This has meant that the Corncrakes of the Western Isles of Scotland and in Ireland have been spared.

Country pursuits

To many people the idea that charging across country on horseback chasing foxes, or shooting Pheasants, might be important ecological factors actually helping bird populations seems bizarre. Indeed one of the species logged as under severe threat is the Grey Partridge and yet the birds are doing best in areas where they are shot! The reason is very simple and, when you think about it, quite obvious. Farmers control the land for farming and they may just think of this as their objective. However many of them have other aims more or less in view. These may be to do with nature conservation or landscape but, over a very large area, they are to do with country sports. The features that are needed for good hunting, for Pheasants and partridges, for hares and for other ground game are also features that assist birds and insects. They are there for the benefit of all the wildlife on the farm but they would often not be there was without the sporting interest – either personally for the farmer or as a valuable sporting let.

In many areas the farm woodlands that remain have clearly been left standing as fox coverts – many are actually called *Fox Covert* – or as places for shooting. Hedges are retained and allowed to grow to a reasonable size. The whole management of the farm has a conservation dimension to it lacking from areas where there is no hunting or shooting interest. The worst farms are those where the top and bottom lines on the farming balance sheet are controlled by the accountant. This is often the case with farms bought some years ago by financial bodies such as pension funds. Farms which are run by the farmers who own them are much more likely to have their objectives softened to include shooting, conservation, landscape values, etc.

All is not well with country pursuits and the practice of hunting, shooting and fishing will undoubtedly be modified either through the pressure of public opinion or through legislation. For instance shoots where tens of thousands of Pheasants are released into an unsympathetic intensive farming environment are pretty awful. The gamekeepers have problems in the crowded release pens and the birds do not provide sporting shots – just a mound of trophies. The local shoot here in Norfolk only releases about 1,000 Pheasants each year onto about 1,500 hectares of excellent varied farmland from large release pens. They have no problem with avian predators as the poults have plenty of cover. The gamekeepers control ground predators and the farm has dozens of pairs of Lapwings, a thriving Stone Curlew population and at least six other species of waders. The small birds like Tree Sparrow, Yellowhammer and Linnet are doing well too. There are, of course, rogue gamekeepers who break the law and destroy protected species but, hopefully, they are a dying breed.

I am personally convinced that there should be better relations between the conservationist and shooters. The amount of money put into conservation – often unwittingly – by country landowners for hunting, shooting and fishing is considerable. Without the personal and pecuniary interest there is a very good chance that many of them would decide that they will join the accountants and run their farming operations to maximise their income. Possibly to use it on extended holidays abroad where they can follow their pastimes within the local law. Once lost, the remaining patchwork of the British countryside would be unlikely to recover without an enormous amount of money being expended over many years from the public purse.

Political considerations

It is impossible to look at the changes to our agricultural environment without considering the political pressures put on the farming community over the years. For instance the reason we have mint sauce with lamb is not just that it tastes nice! In Tudor times when the sheep was most valuable to the country as a provider of wool a law was passed that specified bitter herbs must be eaten with lamb – the least bitter was mint. The idea was to reduce consumption of the young lambs so that they would grow up into nice woolly sheep! I have already mentioned the international trade in agricultural produce and the impact of the two World Wars. I can also recall the startled reaction of the dairy industry when political considerations changed, at a stroke, the support given to the production of milk with no thought of the gestation period of the cow nor the fact that lactations last for many months. More recently the political force which moderates the industry

Yellowhammer.

has been the Common Agricultural Policy (CAP).

The CAP has had fundamental effects and continues to distort farming practice within Britain and Ireland. Systems of support for production have led to farmers farming for the subsidies and thus to the beef and grain mountains and the wine and milk lakes. In turn these led to set-aside to cut down on production of cereals and quota systems for other commodities such as milk. Set-aside (areas that would otherwise be producing crops left unplanted) was to cut down production and not to help the environment. In the first year the instruction was given that the weeds should be cut some time after 15 May. In many cases this was taken as *the* date and so countless ground nesting birds were unnecessarily killed. The date was later changed and environmental considerations were given proper thought.

Now a very small part of the multi-billion pound budget is firmly directed towards wildlife. In specified *Environmentally Sensitive Areas* money is channelled towards farmers entering into agreements for proper management of their land in keeping with its wildlife potential. In other parts – and by no means all – money is available for various types of *Stewardship* which involves the farmers in modifying their actions in favour of wildlife. This may be to make for bigger and better hedges, plant and maintain farm woodlands or even provide weed-rich winter stubbles. Early in 1999 it was hoped that even more would be done when the CAP came up for review but hopes were dashed. The best that could be said is that the reforms were so small that further fundamental changes will have to be made sooner rather than later. In December 1999 the UK Government embraced the arcane option of 'modulation'. A small percentage of the CAP budget is to be put into the agri-environment schemes with additional money from the Treasury. Effectively £1.6 *billion* will become available over the next few years and it is a very real opportunity to stop or reverse some of devastating losses to farmland wildlife. However reform is still needed and the escalating costs of the CAP and the undoubted frauds cannot continue.

However the CAP is not the only, or most powerful, political operation set to mould our countryside. The World Trade Organisation (WTO) seems to be so very powerful that it can overturn national policies and torpedo international agreements. This would decrease the diversity and viability of British and Irish farming and so effect the countryside and our birds.

The future

Our farmland is in a sorry state for its birds and other wildlife. It is not a total desert but it could quickly go that way if there were no changes of direction. The very real problem is that crop monoculture (growing of one type of crop in large areas) is becoming purer and purer so that the amount of food available for the birds is being very much reduced. The introduction of genetically modified (GM) crops has serious implications. These are the GM seeds which confer immunity from powerful herbicides allowing the GM crop to be sprayed and totally cleaned of weeds. In effect, for many species, the green field of crops will be just about as useful as a food resource to birds as the concrete surface of an airport runway. This is simply a part of the evolution of agriculture towards more and more efficient utilisation of the primary productivity from the land by the crop. For farmers who are trying to make a living this is clearly the main concern.

However all is not lost. The public at large benefit from the wildlife of farmland and it seems reasonable for the farmers to receive payment for that part of the management of their land which goes towards the wildlife. This is already happening, to a small extent, through the CAP, but it could be extended greatly. It would be much better for the farmer as the payment would be for doing positive good for our wildlife and not just a cheque for growing, efficiently or otherwise, crops that are not needed by society at large. Payments could be linked to public access, conservation headlands, beetle banks, management of hedges, spring sowing, etc. They could even favour smaller farmers as opposed to the bigger 'barley barons'.

Organic farming may increase, which would benefit the birds and there is likely to be greater emphasis on 'safe foods' – driven by the consumer. This is already happening as the major supermarkets are imposing much more stringent pesticide residue levels on fruit and vegetables for their stores than are needed nationally. It may be that a special 'conservation' class of products which are produced by 'almost' organic methods may come to dominate the market. After all the spread of unleaded petrol was amazingly fast and many people buy free range eggs or have become vegetarian by conviction. Incidentally the further spread of non-meat eating could change the countryside a very great deal. Much of the cereal growth as well as most of the grass growing is for meat production and farming, as we know it, would cease over much of the countryside.

This scenario – farmers ceasing to farm – could come about from market forces as it did earlier this century. This

House Sparrow.

might be excellent for wildlife if the farmers were paid to manage their land for the environmental benefits this could bring. However this could well not be the case and farmers would not want to lock the gate and throw the key away. In most areas scrub and then woodland would become established and farming could then only take place after the trees and bushes had been expensively grubbed out. Most owners would feel that their best bet was to have the land ready for cultivation when 'things got better' and would not take the radical options of conversion to woodland, nature reserve or other leisure pursuits. The horrid vista that might be the outcome is field upon field of brown land as the fields are kept 'clean' by a dose of total herbicide every spring and autumn.

However there is one likely alternative if swathes of the countryside are to be taken out of food production. This is using the farmland to provide energy. Bio-diesel can be made from rapeseed oil, alcohol can be produced from bacterial digestion of all sorts of crops, and the idea of growing short rotation coppice of willows to burn to provide electric power is already being tried. This, in particular, could provide excellent habitat for all sorts of birds and it would definitely change the look of the countryside. These are environmentally friendly means of producing energy as the plants recycle carbon dioxide and do not release the gases locked in fossil fuels. One of the most optimistic events of the last few years was the Government's announcement that the populations of our birds would be one of the 13 ways of measuring the 'Feel good factor'. Now they *have* to do something about reversing the bird declines as their colours have been nailed to the mast!

Greig-Smith, P., Frampton, G. & Hardy T. 1992 *Pesticides, Cereal Farming and the Environment.* HMSO.

Herman, T. 1986 *Seventy Summers.* BBC.

Hoskins, W.G. 1977 *The Making of the English Landscape.* Book Club Associates.

Lack, P. 1992 *Birds on lowland farms.* HMSO for BTO/JNCC/MAFF.

O'Connor, R.J. & Shrubb. M. 1986 *Farming and Birds.* Cambridge University Press.

Open University (with Countryside Commission). 1985 *The Changing Countryside.* Croom Helm.

Pearson, D. 1998 *Breeding Birds of Reydon Grove Farm Suffolk.* English Nature/Suffolk Wildlife Trust.

Rackham, O. 1986 *The History of the Countryside.* J.M. Dent.

Siriwardena, G.M., Baillie, S.R. & Wilson, J.D. 1998 'Variation in survival rates of some British Passerines with respect to their population trends on farmland' *Bird Study*: 45, 276-292.

Whitlock, R. 1983 *The English Farm.* J.M. Dent.

Woodland

Trees are the essence of woodland but woodland is much more than a collection of trees. Forest, copse, spinney, covert, chase, plantation – all sorts of woodland and all meaning special types in different areas. This is the point : woodland habitat is very disparate – this may be because of the different types of tree, the mix of species, the varied management techniques or the different growth regimes in exposed areas compared with sheltered spots. These will also vary all over the country and there are many more species of birds in woodlands in South-eastern England compared with like woods in the far North of Scotland or the very West of Ireland. The rich diversity of our woodland is reflected in the very diverse and rich avifauna that they can hold but we must realise that we are not very well endowed with woodland and that all but a minute fraction – possibly a few ledges on steep cliffs – is heavily influenced by man.

Compared with other habitats, the three-dimensions of woodland, providing up to 20 or 30 metres of depth in which the birds can live, affords them an enormous variety of feeding and nesting niches. Thus, compared with open two-dimensional habitats – such as grassland – there is a far greater variety, number and biomass of birds living in woodland. The trees themselves provide nesting sites for colonial birds which live outside woods – Grey Herons and Rooks for example. They also often provide enormous crops of seed which can be exploited by very large flocks of birds. These can include small finches like Siskins and Redpolls on alder cones and birch seed, tits, Chaffinches and Bramblings on beech mast and Woodpigeons on mast and acorns. The flush of caterpillars on many trees, as the fresh leaves grow in May and early June, is probably the most important feature of our woodland for the majority of woodland breeding birds.

A time without trees

Trees are such an important part of the environment that to be without them would seem unthinkable and yet, about 10,000 years ago, most of Britain was under the ice sheets of the Ice Age and no trees grew. The

Pied Flycatcher.

development of the tree cover spread over Britain and Ireland as the climate warmed and the ice retreated. The succession of different types in different areas can readily be documented from the pollens left behind in peat bogs and it is easy to visualise them. The relevant birds will have spread North and West to exploit the habitat as it became available – they were easily able to fly in to occupy the new expanding habitats as they grew. That is until the huge lake, occupying the North Sea Basin and bounded to the North by glaciers, broke through and gouged out the English Channel. This new stretch of water proved a barrier both to the species of trees that had not colonised Britain and to the more sedentary birds.

The lack of trees and associated lack of structure in the countryside in the Ice Age meant not just no large species in serried ranks of forest but no isolated individual mature trees nor any bushes, shrubs or saplings. No Caledonian pine forest, no beech hangers, no birch-strewn lowland heaths, no alder carr, no lime woods, no hedges, no Welsh oak woods, no trees around the lakes in Killarney and certainly no conifer plantations marching across the uplands! This situation did not change immediately the ice retreated since the soil had to build up fertility and the trees had gradually to pioneer the newly clear areas. Year on year the changes would have seemed very gradual indeed but there were hundreds, even thousands, of years for the woodland to establish itself – and it did eventually clothe the majority of Britain and much of Ireland. Unfortunately this supremacy was not to last too long as the hand of man with flint, metal and fire was about to intervene.

Man and the trees – an uneven match!

The small population of Humans in the country, after the ice retreated in the Stone Age, chipped away at the tree cover with their stone tools and using fire. As the human population increased through the Bronze Age, into the Iron Age and accelerating into the Roman era, more and more woodland was exploited. For instance the industrial exploitation of iron in the Weald of Sussex and Kent consumed huge amounts of wood as fuel, and everywhere wood was needed for domestic fires, for building and for tools. But the forests were cleared not only for the resultant timber but also so that the land could be cultivated. Our woodland might have had a respite from the incursion by man in the Dark Ages, after the Roman era, but this was short lived and the rising population would have increased the pressure. In many areas the woodland character will have been thoroughly changed through use but often the woodland will have been conserved as a valuable resource throughout the last thousand years. This may have been for industrial use before coal became available or, as in the New Forest, as a hunting area for the Sovereign.

Actual instances of destruction are easy to find – the vast wet woodland covering much of the Fenland basin was gradually drained, and then engineered out of existence, largely to become a featureless flat farming landscape. The remains of its former glory can be seen in some of the nature reserves like Woodwalton and Wicken Fen and also where ploughs snag huge 'bog-oaks' from earlier times. These sometimes prove to be as much as 4,000 years old. In much of East Anglia most woodland loss had taken place by the Norman Conquest when many parts of Norfolk, for instance, had many more people living and working in the countryside than at any time since. Here Neolithic, Bronze and Iron Age and Roman remains can be found all over the county save only those areas that would have been very wet. Far from the flatlands of the East, in a very different habitat, the remote uplands of Scotland are often eroded peat with the stumps of the old Scots pines exposed. These are not generally as old as the fenland oaks but they were just as surely killed by Man's activity. In this case the open pine forests of the Highlands were a valuable resource and were systematically destroyed by the feuding clans during the Mediaeval period and, later, by the invading English. Even as late as the 1960s some areas of the old forest were being felled to make way for plantations. Now about 98% of the ancient Caledonian pine forests has been lost. Enough remains to show how well adapted was the mix of trees and the varieties of the pines to match the local conditions. In Ireland native forest seems to have held up well to about 1600 when about 12% of the land area was still covered; this was down to about 2% by 1800. These clearances may have rendered extinct some forest species of birds – like the Great Spotted Woodpecker and Capercaillie.

This tide of destruction was really very thorough and left Britain and Ireland with less forest and woodland than any other European country – down to 5% of land area in Britain by 1919. Then the economic danger of this situation was realised and, with the founding of the Forestry Commission, much planting of woodland took place. Currently Britain has about 10% woodland – well below the average for Europe of 25% and only Ireland (6%) has less. The ecological effects of 'management' were also very far reaching. The initial resource was exploited in many ways, from the felling of ancient trees for structural wood in cathedral roofs and the state's use of large trees in warship building down to the legal use of fallen dead-wood by local cottagers for fuel. Indeed in many areas there were rights to gather gorse for burning in bread ovens! Even activities not strictly to do with the trees had, and have, a devastating effect on regeneration. The right of 'pannage', running pigs in the woods to eat the acorns and beech mast, will have inhibited the growth of young trees and any sort of understorey. Running pigs in woodland is not now usual but huge tracts are very heavily grazed by sheep and many oak woods in Wales, for instance, have only had new growth (regeneration) very infrequently in the past. Their open aspect is wholly unnatural as the stocking level, by the farmers of sheep, far exceeds any grazing pressure that might happen in nature.

Natural woodland

Our natural woodland must have been marvellous: rich, diverse, extensive, but now all gone. Even the areas of ancient semi-natural woodland that we quite rightly value, like the New Forest, Burnham Beeches and the Forest of Dean, are pale shadows of what they should be as they have had trees selectively harvested for many years. The regeneration which would have happened has suffered terribly from a wholly unnatural grazing pressure. In Europe it is only really in the primeval forests of Eastern Poland that the full glory of lowland natural forest can be appreciated. Here the trees are very much taller than any but the most exceptional in Britain. The conifers in the mixed woodland poke through the top of the canopy and provide natural nest sites for Swifts in their snags. Wind throw of the biggest trees tears them out of the ground and creates vertical root discs six or eight metres high. At first these are bare earth but are quickly clothed in a rich mix of plants. These provided valuable nest sites and the resultant clearings could be big enough to accommodate territories of many species. These are often ones that do very well in the normal, impoverished, woodland situation we have in Britain – Dunnock, Robin, Blackbird, Song Thrush for example. The Polish species range of trees and birds exceeds that which we would ever have had in Britain but the structure of the woodland would have been similar.

Even lowland forest areas would have varied according to geology, hydrology and climate. Lowland wet areas with reeds would have become clogged with vegetation; willows and alders start off the woodland but, as it dries out, oak would have been the climax. In better drained areas the alders would have been confined to the stream sides and the main trees might have been beech, oak, elm and hornbeam with birch and rowan infiltrating the gaps where wind-throw had allowed in more light. On very light soils there may have been little more than a light covering of birch and rowan with a few large, slow-growing Scots pine on the drier areas where soil had accumulated. In many other situations the native small-leaved lime would have been dominant – as it still is in parts of Lincolnshire – or woodland might have contained many ash trees. Holly, juniper, yew, hawthorn and willow all had their part to play but, even though they are still present in wood and scrub, they are a part of a complex succession interrupted by man's activity and never able to reach its climax.

This disruption does not alter the value to birds and wildlife of the habitats even 'damaged' woodland presents. Indeed the changes wrought by man have often produced a situation where a rich assemblage of plants

Woodcock.

Coppiced woodland.

and animals can find good conditions – look at the bluebells, Red Kites, Pied Flycatchers and Wood Warblers of the Welsh sessile oak or the oxlips and Nightingales of the Suffolk coppiced woodland. However both the habitats require management to retain their interest. The oak woods will alter in character if grazing is restricted as saplings create a dense shrub layer, but, unless grazing is restricted every hundred or two hundred years the oaks will fall over and the woodland will be without trees. In the coppiced woods the structure would be lost without the active cutting of the stools. In the coppiced woods the structure would be lost without the active cutting of the stools. The light would be excluded from the woodland floor and the plants would remain dormant and the lush ground layer – needed by the Nightingales – will be lost. Providing a nearer approximation to 'natural' conditions, uninfluenced by man or his animals, would be an interesting experiment but often an unsound one for the areas involved are probably much too small to allow the natural processes to work properly – and the effects of pollution through acid rain and global warming might upset the experiment anyway.

Managed woodland

This describes our existing woodland. The trees in Regents Park or in the suburban gardens of any city are clearly heavily managed but still have important elements for birds – try asking the breeding Grey Herons on the islands in the lake in Regent's Park! Clearly the open parkland, so much associated with the great houses and such names as Repton and Capability Brown, are a living embodiment of a romantic dream and very different from anything really natural. Many people familiar with the lovely beech hangers of the Chilterns, and other chalky areas, will be very surprised to know that these are not natural but are the result of very careful planting and husbandry. Indeed it seems that many of them were planted in a series of pits dug in the chalk and filled with top-soil so that the trees could thrive! The management of natural woodland alters its character in many different ways. The crop will always be taken away depriving the ground of the materials from the decaying wood and denying many fungi, bacteria and insects their potential homes. Particular species will be favoured and those not wanted thinned at an early age – and the trees being cropped are often taken very young. Indeed in many cases it may be a shrub for which the wood is managed – hazel in areas where hurdles are important – and the only large trees are the few standards left. In many areas the natural tree cover was largely destroyed to be replaced by more valuable crops like sweet chestnut for paling fences.

These changes in tree species were also accompanied by huge changes in the structure of the woodland. Looking at the map and seeing green woodland marked in an area which would have been so marked for hundreds, even thousands, of years does not mean that it will be anything like natural – or the same – just because it is old. Over the years the use of the woodland will have changed with ownership, with the changing needs over history and with the tensions between the various interests using the area. For instance the rights to take firewood or graze animals will have been an important pressure on many woods until the end of the 18th century. Then the sporting interest in the woodland may have led to changes in the structure – rides for the guns, shrub cover for the Pheasants, etc. Now the woodland may be the subject of grant aid which carries with it the need to take care of wildlife and allow public access.

These changes from the 'natural' structure are very important for the birds. Two ancient methods of management involved rather frequent harvests from trees – without killing them – and gave rise to two different sorts of regeneration. Coppicing, where the crop was taken from ground level, was favoured in many areas. The continuous supply of a crop over an extensive area of woodland was often very important and the produce from several small 'parcels' within it might be sold each year on a 12, 15 or 20 year cycle. This would mean that the birds had a mosaic of different aged habitats in which they could live. Tree Pipits might like the first five years and Nightingales years four to nine but each would have the possibility of breeding within a quarter of the total area in any one year without having to move far. The vogue for coppicing small remnants of woodland for conservation purposes has many uses but is not particularly bird friendly as the area involved is so small. Bigger woods where two or three hectares can be done each year are much better. The other cropping method was not nearly so good for birds as it allowed for grazing by animals within the 'wood pasture'. The new growths were taken, in rotation like coppicing, from eight or ten feet above the ground where the new shoots would be safe from the teeth of the animals. This is known as pollarding and was in use in many areas. It is particularly associated with willows in wet areas where the canes may be cut every two or three years.

A very special sort of managed woodland is the orchard. These have played a very important part in the countryside for many centuries and vary from a few trees associated with a garden, through an acre or two associated with a large house or farm to the serried ranks of fruit trees to be found in Kent or the Vale of Evesham. Many of the trees used to be large and old but more and more of the diminishing area of orchards now consists of much smaller trees. This is largely through the modern realisation that harvesting fruit from bigger trees is seriously dangerous. Many commercial orchards currently have a good deal of pesticide use and may have few insects for breeding species, but apple orchards will often have huge amounts of waste fruit available for wintering birds.

Tree planting

Much of the woodland that we now have in Britain and Ireland is a planted crop of softwood (conifer) being reared for harvesting. Often this is private planting on otherwise unproductive ground and the economics of the whole enterprise was often based on grants and tax incentives. In many areas the woods are doing well and will provide a decent return – in terms of money and wood – but in others the planning, research and husbandry has been sorely flawed. Some woodland has been planted in such remote areas that the cost of bringing out the crop would exceed its value. In other places exotic species have been planted and have proved to be wrong for the climate and soils. These are practical, forest oriented, mistakes but further problems – or perceived problems – have been associated with much planting. Many of these do not now exist both because regulations and tax systems have changed and there is more understanding of landscape and wildlife considerations. For instance there is little, if any, new planting of conifers in lowland English areas where they were inappropriate. In Ireland ill-advised planting of peatland areas has tailed off and the pressure on the Flow Country in the very North of Scotland has been relieved. Where planting does continue

Goshawk.

it now generally follows the natural lines in the countryside rather than carving it up with straight boundaries.

Many upland areas in Britain were woodland, and the moorland, being lost to forestry, was itself a very degraded habitat. However often there are populations of rare and interesting birds, notably waders and raptors, living there which are lost when the trees grow. Indeed virtually all the birds of the moorland areas will be displaced; this does not mean that the new plantations become bird-less – in fact there will often be a great deal greater biomass of species in the trees than on the moors. It is just that the Meadow Pipits and Skylarks are replaced by Chaffinches, Coal Tits, Siskins and Crossbills, scattered Red Grouse by Woodpigeons and so on. Modern plans for forests often leave areas unplanted beside streams and in areas where the ground is especially steep. These sites are exceptionally valuable for plants and birds, and can retain all sorts of species – from the streamside Grey Wagtails, Dippers and Common Sandpipers to the Hen Harriers. The density of the planted trees is designed to maximise the production of timber and there are very many more than would occur in any natural situation where, in any case, one would expect several species and not a mono-culture. The rather small areas unplanted are thus very important in allowing birds that are not adapted to the pure stands of conifers to breed.

Now that many forests are being harvested we can have a proper succession. This is much better for the birds than vast areas of woodland of the same age where scrubland birds do very well for the first few years but they are then wholly excluded and are unable to exploit the first replanted areas. This has proved really good in Thetford Forest where there will be newly planted and developing areas for Woodlarks and Nightjars, scattered over the whole forest every year, whilst there will also be big trees in other areas where the Goshawks can breed. There have been sales of smaller areas of woodland outside the main forests as the economics of working them properly has been seen to be very poor. In some cases these sites are actually restored to heathland or are planted with broad-leaved trees for their amenity value.

Not all new plantings of trees are in massive conifer plantations. Trees are being planted in country parks and nature reserves, gardens and along motorways and, in the farmland context, along hedges and in farm woodlands. A really exciting new initiative is in the plans for national forests which will be mosaics of woodlands covering large areas of England. Some planting has started and the planning for these is in place, but it remains to be seen whether they really do come to fruition. These are very long term plans but the modern countryside is also seeing the ultimate in short term woodland! This is the use of willows on a two or three year 'short rotation coppice' management system for the production of fuel – not much good for traditional woodland birds but probably truly excellent for such species as Sedge Warblers.

Some woodland problems

All is not well, at the moment, for our woodlands and their birds and other wildlife. The pressures on owners and managers from the insurance companies to ensure their woods are 'safe' is having a very detrimental effect in some areas. Ordinary accidental injury and damage insurance is invalidated if sick or dangerous trees are not made safe and have subsequently fallen and caused a claim. This has the effect of causing large numbers of trees in public places and along roads and tracks to be felled or mutilated to avoid such accidents – even if there is only the very remotest possibility of it happening. This is a particularly insidious problem as the very act of asking for expert advice implies that the owner is not satisfied that the tree or trees are safe. The expert providing the advice might be liable if an accident does happen and must be under pressure to suggest that felling or remedial work should take place as the prudent option. Old trees with holes large enough for bigger birds such as Tawny Owls to nest in are certainly very likely to suffer and dead trees are a very valuable habitat for breeding insects.

The health of our trees is also a real problem. The virulent form of Dutch Elm disease, imported from Canada in consignments of felled timber has had a terrible effect on the numbers of elms in many areas and devastated the farmland landscape in areas like the Constable country of Suffolk. There are also diseases affecting such species as alder, oak and beech. Insect pests have been imported with dire effects – both on native species and also on exotics. The effect of acid rain and other pollution has devastated some woodland but has destroyed many trees in farmland and

Grey Herons at heronry.

there are areas where it is rare to see a farmland tree without the stag-headed tell-tale of die back. Even the ecology of our woodlands where no domestic animals graze is being altered by the escalating numbers of deer – particularly the roe and muntjac.

One perverse effect of supposedly good management – the deer fence – has been appreciated by many managers. These tall fences were erected to save areas from the grazing pressure of the deer but had the apparently trivial side effect of killing the woodland grouse – used to being able to blast their way with impunity through the twigs of the local trees. It was not realised, for many years, that the occasional body found represented a very significant new mortality factor for the Black Grouse and Capercaillie. The idea of allowing natural regeneration is fine but there seems little option but to control the deer more rigorously. Thankfully grant-aided schemes are now available to assist with the removal of the fences in many areas. Similarly some totally inappropriate plantings with exotic conifers have been removed from sensitive sites before the trees have reached maturity and started to seed themselves.

The future

Recent weather events have often had dire effects on our woodland. Many million trees were lost in the South-east of England in the hurricane of 1989 and there have been many other 'great storms' recently. It is amongst the predictions of the effects of global warming that there will be more violent weather as well as a greater incidence of flooding and drought. The former can cause immediate losses of trees from water-logging and land-slide but the latter is probably the most worrying. Many trees of a wide variety of species have died in periods of dry weather through water-stress and it may be that the native species of trees will be found to be unsuited to the natural conditions of the future. This weather-induced damage may often trigger a wish to 'clear things up' and this has to be done to salvage a crop from wind-blown plantations, but semi-natural woodland is best left alone, as far as possible, to let the trees come back naturally.

However the biggest change in the future is likely to be an increasing public perception of the recreational and cultural value of woods and trees as well as being a very important wildlife resource. Many of the road building protesters have concentrated on woodland and, typically, build tree houses and try to defy the developers. They do not occupy barley or wheat fields and planners would be well advised to take this into account: natural and semi-natural sites like woods and marshes have an enhanced social value. It is easy to come up with a reasonable eco-nomic compensation for building roads, factories or houses on an area of farmland. Its value, as farmland, is known and there is little hidden social value and, at the moment, we do not need the food production for us to remain viable as a nation. In addition there is the move, over the last century, to make woods and forests into nature reserves of all sorts – these should be safe for all time rather than at the whim of the owner's need for money.

There is, however, a rather dismal aspect to modern life as land becomes more and more valuable. There are many Victorian areas where the developments of housing were of such low density that their large gardens, planted with trees, have developed into a fairly good approximation of woodland. Very few modern developments have any chance of such a green outcome. Even where trees and shrubs are incorporated in the developments, they are often plantings of exotic species and are planned to be carefully manicured and kept in check.

Avery, M. & Leslie R. 1990 **Birds and Forestry.** T & AD Poyser.
Fuller, R.J. 1995 **Bird life of woodland and forest.** Cambridge University Press.
Rackham, O. 1990 **Trees and Woodland in the British Landscape** (revised edition) J.M. Dent.
Steven, H.M. & Carlisle, A. 1959 **The Native Pinewoods of Scotland.** Oliver & Boyd.
Yapp, W.B. 1962 **Birds and Woods.** Oxford University Press.

UPLANDS – MOORS & HEATHS

'Uplands' in Britain and Ireland may not be too far up! In many countries even our tallest mountains would hardly count as a hill. There is a fairly clear distinction between ordinary farmland and the areas which can be called 'upland'. Conventionally this is supposed to be over 1,000 feet but this is not a useful distinction in habitat terms. Almost all land over this height is certainly upland but many marginal areas, a lot lower down, should also be included. Possibly we should be looking at a different definition which might be based on the hill subsidy system of the Common Agricultural Policy! 'In by' land, in Scottish terms, would not be included. These are enclosed fields which may be used for grazing, hay, silage or arable. So for our purposes 'upland' is generally higher but never arable and often un-enclosed. Where there are fields they are usually divided by stone dykes and not hedges. The land is usually used for 'rough' grazing, forestry or simply be mountain and moorland 'waste'.

However we have to be careful for the definition from the CAP would leave out huge areas of the lowlands which fit best in this chapter. Again these are mostly used for low level grazing, for peat cutting or not used at all. These include lowland heathlands like those of Suffolk, Surrey and the New Forest where they have not been ploughed, become wooded or deliberately planted with trees. The South Downs, the Chilterns and Breckland include open areas, well below the 1,000 foot mark, which are included in this chapter. There are many heather or bracken dominated areas, all over the four nations, which qualify whether or not they are 'upland'. Vast areas of Ireland are not 'up' but whether they are peatlands or the limestone pavements of the Burren, they belong here. Extensive grasslands have also to be put in this category. So this chapter treats the arctic wastes of the High Tops of Cairngorm and the other Scottish mountains downwards! These habitats seem to have low human inputs but this is not always the case and several have been shaped by man in the past. These tend to be the places where 'Right to Roam' confrontations are going to take place over the next year or two. This is particularly the case when the open habitats appear as fragments within the intensively farmed landscape or are used for sports such as grouse shooting or deer stalking.

The top habitats

It is a different world on the tops of the highest mountains, and down below 1,000 feet in the North-west Highlands. These are almost arctic areas, which does not mean to say that they are incredibly cold during the summer or that they are lacking in interest. Indeed they are often rich in flowers and have a very special range of breeding birds – some of which are not found nesting anywhere else in Britain. They are apparently barren and remote areas but even here they are under threat from man. The Cairngorm plateau has ski lifts and a restaurant up to and on the most fragile of habitats and it is literally worn out. This exploitation of a transient resource – snow slopes – is a continuing problem particularly as the interest in skiing is increasing and the investment in the resort, and therefore the number of visitors, is increasing. Much of the site is properly protected – SSSI, NNR etc. – but this does not stop planning application after planning application being filed. Each time the destruction of habitat, once and for all, is at stake and the conservation bodies *have* to defend it. This costs money, for both sides, but victory for the developer would actually be very profitable, whereas victory for the conservation interest is simply a job well done and money spent. There should be a statutory ban on such multiple serial applications which are totally exhausting to the conservationists. This very special habitat is, naturally, slow to change and easily damaged just by feet so that the effect of the visitors who walk around the top of a ski lift will be pretty devastating.

Even away from areas of easy access, walkers and climbers can disturb rare birds – especially if climbing routes were to target the crag where a Golden Eagle or Peregrine was nesting. However other birds of the tops, like Ptarmigan and Dotterel, are not particularly wary of man. The localities of the very rarest of the species that breed in these habitats are close kept secrets and truly astonishing birds can turn up and apparently consider breeding – like the Long-tailed Skua sometime ago! Unfortunately, just like other species, these birds do need two, of different sexes, to breed successfully! There are very large areas of this habitat which do not have large numbers of birds breeding but some, like the Golden Eagles, have and need huge home ranges. Other species are spread thinly over the whole area and still others have special requirements. For instance there are dozens of pairs of Snow Buntings which rely on craggy areas for nesting and summer snow fields for feeding – not on the snow but on the wind-borne insects that show up on its surface.

The topmost habitats contain large areas where much of the ground is hardly vegetated and the wind and winter snow are clearly the most important shapers of the landscape. They are above the tree line – the altitude up to which real trees will grow although one can find dwarf species within the ground vegetation. Further down there is more vegetation – heather, various dwarf berry- bearing shrubs and grass – and a greater range of birds. Common species, like Skylark and Meadow Pipit, and waders, like Golden Plover and Dunlin, are typical species. These areas will probably be grazed by animals such as red deer and sheep and not just mountain hares. The Ptarmigan will give way to Red Grouse but it is some way further down the hill that you will find the areas actually managed as grouse moors. The high moorland areas are not immune from over-grazing but the slightly lower areas are more likely to suffer and may, in some places, also be 'improved'. Unfortunately improvement does not improve the bio-diversity of the sward nor the populations of breeding birds. New varieties of grass, new machinery and modern drainage methods have meant the steady march upwards of the bright green of re-seeded areas. This was heavily subsidised and mainly happened some years ago. The subsidies have changed and there has been very little money in rearing sheep in recent years. Even lower, where cattle might have been raised, the curse of BSE has caused huge disruption.

Hundreds of thousands of acres of these habitats have also been lost to forestry. In some cases this has come about because of tax incentives for forestry. Proper survey work would have shown that these should never have been planted as the ground will not, cannot, support trees. This may be because it is too impoverished, too wet or too exposed. However large areas are now growing trees and the subsequent development of the woodland is dealt with in the appropriate chapter. What few people realise is that much British moorland is actually the scene of desperate desolation brought about by human activity. These areas should not be moorland heather but open woodland which, over much of Scotland, would often have been dominated by Scots pine and birch. The Caledonian pine forest seen at its best in Speyside and some parts of Glen Affric would have covered much of the area and the stumps are often still there buried in the peat. The wood was a useful resource so the feuding Highlanders would burn and destroy it to deny their enemies its use. This has serious consequences for many areas in the uplands. They are living on the fertility built up by the few thousand years when the woodland thrived since the last glaciation.

One of the highest profile uses of these moorlands is for the traditional driven shooting of Red Grouse – opening on the 'glorious' twelfth of August. The grouse moors are managed for the grouse by a mosaic of burning which provides them with new growth heather shoots to eat. The birds are susceptible to parasites which naturally cause the population to cycle with booms every four years or so – in these years a couple of thousand grouse might be shot off a moor that, in other years, will provide two hundred or less. This cycle can be beaten if a good proportion of the adult grouse are caught in the spring and dosed against the parasites. So far so good. However the fertility and productivity of the moors may well be dependant on the nutrients locked into the system when it was a natural, open forest. The grouse moor management regime may be viable for a century or more and then it could fall apart and this may be one of the reasons for consistently poor bags in some areas. In many areas illegal persecution of birds of prey – particularly Hen Harriers and Peregrines – goes on in a clandestine way as the keepers see this as the only way to get good numbers of grouse in front of the guns.

A recent research project on the effect of birds of prey on grouse showed that on the Earl of Dalkeith's moor at Langholm (South Scotland) the Hen Harriers and Peregrines increased when the keepers protected the raptors. They had a pretty disastrous effect on the Red Grouse so that the expected peak of birds for shooting – about 2,000 – failed to appear. However the obvious deduction may not be right : on a Highland moor protection did not cause a big increase in raptors and the bag of grouse peaked as expected. The difference in ecology between the two places would seem to be the cause. The Highland moor is dominated by heather and has some red deer and few (if any) sheep, but Langholm has lost much of its heather to grass and is heavily grazed by sheep. This has meant that the numbers of small birds, like Meadow Pipits, has increased and they are the food the Hen Harrier males need in the spring. Their numbers have increased and later in the season they can catch and feed on young Red Grouse because of the lack of heather. Young grouse can easily conceal themselves under heather but the bird of prey, flying over, can see the chicks down the vertical stalks of the grass. Sadly, driven grouse shooting has become uneconomic on Langholm but steps are being taken to reduce the sheep and encourage the heather so it may eventually be possible again in the future. This would be good news for many other species and not just the Red Grouse.

The political situation in Britain and Ireland – and latterly in Europe – has shaped the face of the uplands just as much as it has the more intensive farming environment. For instance the support for sheep farming, and to a lesser extent cattle, has been based on the individual sheep and cow and not on the farmer or the land unit. This has put a premium on 'efficiency' and the stocking rate has increased so that there have been terrible problems, concealed from most of the public, with welfare. In particular there are very many ewes who die in the winter and early spring on the hills. This has been good for carrion-eating birds like Ravens, Buzzards, Red Kites in

Golden Eagle.

Wales and Golden Eagles in Scotland. It has also led to the loss of the smaller hill farms and a characteristic of almost all marginal areas is that there are many abandoned dwellings. It has also compounded the problems for the habitat not just through over-grazing but by increased sales of the land for forestry. The new farming economics have forced out the generations of small farmers that used live in the hills.

Loss of the land to forestry might be seen, from what I have written earlier, to be a very good thing as so much of the lower areas within the uplands used to be woodland. However this is not the case as the regimented ranks of conifers, almost always of exotic species, marching across the rolling hills are very different from the open and diverse Caledonian woodland and the similar forests that clothed the moors. Further discussion of what happens in the remnants of the old and the regiments of the new woodland appear in the forestry chapter. The sort of upland birds that have been lost from the planted moorland concern us here. Clearly huge numbers of such species as Skylark and Meadow Pipit, that will only nest on open land, have no chance of finding a living in the woods. Waders such as Lapwing and Curlew will also be lost but the stream-side Common Sandpipers are probably able to hold on as are the Dippers and Grey Wagtail. The streams might become acidified so that the invertebrates are lost and with them the birds disappear. Clearly

47

the Red Grouse will also go almost immediately and gradually, over the first few years of the trees growing, birds like Cuckoo, Hen Harrier and Merlin will be lost from dense plantings. More recently the idea of 100% planting has, thankfully gone out of fashion, and the unplanted catchments of the streams are likely to retain these birds.

Not so upland!

As I have explained, some 'upland' habitats are actually rather low down and very much threatened. These include the lowland heathlands characterised in south-east England mainly by the fragments left in Surrey, Suffolk, Norfolk, Hampshire and Dorset. Here the pressure has come from agricultural development, from afforestation (both by planting and by encroachment), from sandpits and even by development for building and roads. By some accounts we have lost 90% of the heaths that existed before the Second World War! Many of these heaths are considered to be very important with scarce bird communities breeding on them and such rare animals as the smooth snake and sand lizard. Much of the area is now protected as Nature Reserves or by designation as SSSIs. Despite such designation further fragmentation has taken place in recent years especially through building and, because of earlier permissions granted many years ago, quarrying. In recent years conservation bodies like the RSPB and English Nature, together with the Forestry Commission, have started to restore heathlands. These have concentrated on areas where the trees have recently been harvested but, still, there is probably a

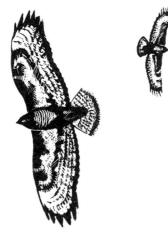

Buzzard.

considerable net loss annually of this habitat. However one does have to concede that the tiny Dartford Warbler, once very rare and at risk from cold winters, is bucking the trend and doing very well at present. It has even made the jump across the Thames Estuary and colonised the coastal areas of Suffolk! This sort of habitat, with light hawthorn scrub, was where the last of the British population of Red-backed Shrikes bred and these, sadly, seem to have gone for good – or at least for many years.

These areas are generally very

sandy and were used by the local people for grazing and for wood gathering. The latter was an activity which severely restricted the tree cover growth. The rights vested in ownership or tenancy of particular properties was generally prescribed in various ways: in some woodland they were only allowed to gather fallen branches, in others gathering could only take place in the winter but in some areas the wood gathering was unrestricted. One 18th century report on a common in Hertfordshire stated that even six inch stems of gorse were being collected to fire the bread ovens and there were no trees. Grazing was by herds of cattle, sheep or goats which were tended by small boys and shepherds and brought back to their pounds overnight. Here the dung was collected and used to manure the arable areas – and, because of this, the heaths were further impoverished. One expanding feature of many of these areas is the coverage of bracken and this is likely to continue. This invasive plant may take over as much as twice as much of the countryside in the next 50 or 100 years due to global warming and increased carbon dioxide. It is a very useful nesting habitat but chokes out other plants and does not have the insect fauna that they would have and the birds need.

In the areas dominated by chalk – like the Downs and the Chilterns – the close cropped turf, areas of long grass, hawthorn and even juniper scrub used to be a common habitat peopled by shepherds and their flocks. This was impoverished by the animals being taken home at night and their dung collected. Now such habitats have retreated to a very small part of their previous extent and mostly to the steeper parts of the hills. Where stock is still kept it stays out at night and the dung is not gathered. Most of the rest has been ploughed and decent crops of cereals, linseed, oilseed rape and sugar-beet are grown – supported by the CAP. There have even been recent cases where areas supposedly protected by designation as an SSSI have been ploughed for cropping with linseed! In one case there was uproar! The Secretary of State issued a section 29 order and the farmer stopped. An army of volunteers replaced the ploughed turf and there was a small, but significant, victory for conservation. Loss of this habitat has probably had a worse effect on the botanic diversity than on birds but the losses will have been bad. Losers will have included such species as the two whitethroats and Linnet. These species will actually have been helped by another factor which has caused the grassland habitat to be taken over by scrub – myxomatosis. The loss of rabbit grazing, taking out trees and shrubs before they have a chance to establish themselves, has greatly altered the habitat since the late 1950s. It has also caused the popular myth that conservationists spend their time doing nothing but chopping down scrub!

Another lowland habitat in this category is the breckland of Norfolk and Suffolk. This area of poor soils over sand and chalk used to have very extensive areas of lowland heathland. Over the last 200 years techniques of farming and the planting of wind breaks, traditionally of Scots pine, have brought some areas under cultivation. More recently very large areas have been planted for forestry (see Woodland chapter) and these have mostly been successful. Once more grazing shaped the area over many centuries and the presence of the Stamford battle area, grazed by thousands of sheep, has probably saved some areas which would otherwise have been lost. On several areas there has been public outcry at attempts to return commons to their proper, open, habitat. The local dog walkers like the trees which have invaded in the last two or three decades and they think this is what the area should look like. A few moments' consideration of the typical breckland birds – like Stone Curlew, Woodlark, Nightjar and the locally extinct Red-backed Shrike – points to heathland as being the 'proper' habitat. This rather small area holds some of the richest areas for native plants in Britain. It is now designated as an ESA (Environmentally Sensitive Area) so that money from the CAP does really go to conservation.

In Ireland the major cause of the depopulation that created much of the 'upland' habitat from areas which were, until the middle of the 19th century, farmed habitat, was the potato famine and the subsequent emigration to America.

Another of these non-upland but undeveloped habitats is the very varied land immediately beside the sea but not really a part of the coastal habitat. Whilst modified by the close presence of the sea and its wind and salt spray these habitats tend to be very like the local 'uplands' because of the underlying geology – the chalky areas above the Seven Sisters in Sussex is not really like the equivalent area on the cliffs of Pembrokeshire. However both are rather warmer, through the influence of the sea, than the equivalent habitats inland. This means that they are rather better places for resident species to live in when the weather is severe during the winter. Thus they are, for instance, good places for Stonechats.

The future

These areas are liable to rather swift and radical change. Much has already been lost to forestry, for example, but it is interesting to speculate what will happen as these come to be felled. It is possible that they will produce a single good crop in some areas, possibly two or even three in others, but it is likely that many will not be worth replanting after two or three crops have been taken. As this happens clearly there will not be money readily available from the forestry interests to restore the land to what it was before. What will happen then? We all realise that times are very hard for farmers who are trying to rear sheep and that damage is being done to the uplands as the stocking rate increases. On the other hand much of the area we are talking about needs some grazing and the uplands would change quickly were the sheep to go. The preferred option, amongst conservationists, is that the farmers should continue to receive support but that the coupling of this to numbers of sheep on the hill should be broken.

This really begs the question of the *value* of the uplands for the owners, for the rest of society and for wildlife. The owner may only be able to make a small annual income from the sheep on his land. However, for 'society', the contribution the rather small parcels of land make to the total 'wilderness' area of the local National Park is much greater. This is enjoyed by hundreds of thousands of visitors and the actual monetary input into the local economy is vastly more than the small amount made by the owner from the sheep. In wildlife terms there is a (small 'c') conservative feeling that the best use of the land for wildlife is to keep it as it used to be – or to restore it to what it was. This may mean that conservation bodies, who have already bought quite large areas, change the face of the hills and return them to something which is very different from what there is now. For instance in mid-Wales a sizeable tract of bracken covered sheep walk is being planted with hundreds of thousands of local oak trees to extend the oak woods. Mostly these only remain on steeper slopes. Such imaginative schemes are likely to become more general.

One thing is certain. It has been demonstrated time after time that the different methods of exploiting upland cannot co-exist. A grouse moor, to be successful, needs to managed as a grouse moor and there is little chance of anything but very low density sheep rearing. Sheep and/ or Red Grouse cannot co-exist with plantation forestry. Forestry and high-density deer herds are a recipe for disaster. This means that very careful decisions, long-term, have to be taken and held on to over quite long periods for the future. Clearly the uplands are much too important for their future to be left to chance. There are many species of birds and plants, often rare and important, which depend on these areas and hundreds of thousands of people who live in or choose to visit these areas. The right to roam, the issue of absentee landlords, the economic problems of farming and the problems of forestry and peat cutting all affect them. Unless we get the answers right many opportunities are going to be lost over the next ten or twenty years and then it may be too late.

Ratcliffe, D. 1990 *Bird life of mountain and upland.* Cambridge University Press.
Redpath, S.M. & Thirgood, S.J. 1997 *Birds of prey and red grouse.* London: Stationery Office.

WETLANDS

For a large number of birdwatchers their local honeypot for birding is a reservoir, gravel pit or marsh. These freshwater habitats in the countryside attract a wide range of birds and many species not found in other habitats. However the history of wetlands within Britain and Ireland is very mixed and many of the sites that we go to now are recent additions to the landscape. Not only that but they are also artificial sites which have come into existence, deliberately or by mistake, as a result of human activities. The lack of old wetland habitats, apart from flowing water, is easily explained: old sites tend to have been drained in the past for agricultural use or they have simply silted up and dried out. Indeed geologists look upon wetland areas as being temporary and almost always subject to gross change over what, to them, is the very short term. This chapter will look at the whole range from wet meadows to huge reservoirs and lochs, from babbling brooks to wide and sluggish rivers nearing the sea.

All sorts of wetland will have been around before man started to shape the landscape. As the ice retreated, following the last glaciation, the rivers and streams will have followed the line of least resistance down to the sea. Where the ice had scraped out sumps or laid down damming moraines, water gathered. Ponds and lakes were formed, creating open water and marginal habitats. However as the vegetation grew up many of the open water habitats became swampy and even dried out. In Norfolk the pingoes – formed from small depressions in the glacial moraines – have remained healthy and wet for thousands of years but they are an exception. In some areas, particularly in Ireland, limestone areas flood in the winter as the water table rises forming turloughs. In some coastal areas sand or shingle bars have formed and bank up fresh (or brackish) lakes. These are again rare wetlands that may have stayed with us for many centuries.

However look at any map of lowland Britain and Ireland and most of the areas of blue, showing water, are recent lakes, reservoirs, ponds or pits. They are often in areas that would have been wet in earlier times – it is those places where the water tends to accumulate and most sand

Moorhen.

and gravel pits are in areas where alluvial deposits are available for mining. However even such vast wetlands as the Fen Basin, south of the Wash, where a map a thousand years ago would have shown a massive and continuous marshland, is now almost wholly dry. Indeed it has also shrunk as the wind has lifted the sand and peat off the drained land. Hundreds of square miles of marsh have disappeared. Old maps, here and elsewhere, show by the pattern of field boundaries how the wetlands have been drained, while ghostly patterns of banks and drains indicate where some very interesting bird and natural history sites used to exist.

Fens, bogs and marshes

These are just as much wetlands as areas of open water and, in the lowlands, equally likely to have been drained and transferred into agricultural production. The majority of these areas are associated with water-courses and the regular flooding and spates even now serve to keep them wet. In many cases the habitats are very likely to deteriorate gradually and dry out – often becoming woodland if man or flooding does not intervene. However man does very often intervene and drainage schemes have been evident for centuries – even millennia! For instance villages on the Somerset Levels were actually built on stilts to keep them above the flood levels when the Bronze Age people were living in the natural habitat. Now these lake villages and timber pathways are being dug by archaeologists on reasonably dry land – reclaimed centuries ago.

As one approaches the final bank between the sea (the Wash) and the land in Norfolk and Lincolnshire one sees younger and younger drainage banks. The innermost are often called 'Roman Bank' and there are salt pans certainly dating back to Roman times. The villages often have the same name – but with a different Saint attached. These indicate newly reclaimed lands which were, at first, attached to the existing village until they became big enough to have their own church. Then the name stuck but the village was named after the new church. For instance the old town of Holbeach has Holbeach St Marks and Holbeach St Matthew to the north of the town in the old marshland. The rich, flat fenland area with its peat and silt soils is now fairly featureless and home to few birds. Before the drainers came, particularly the Dutch some 300 years ago, this must have been the most incredible landscape and there were certainly such birds as Ruff, Bittern, pelican, Spoonbill and Crane present. Possibly the Danube Delta might be the nearest habitat within Europe with which comparison could sensibly be made.

So what is there left of wetlands in the Fens? Almost nothing! There are the rivers, of course, and a few pits from which the material for the banks, or even gravel for construction, has been extracted. However the really good sites are up in the air! Whilst the ground has eroded and shrunk around them the areas used to store water in winter so that the floods do not spread are several metres *above* the surrounding land. These marvellous sites are the washes – particularly the Ouse Washes. They are about 30 km long and can have over 100,000 wildfowl on them in a good winter – with Wigeon and Bewick's Swans (*Cygnus columbianus*) as big players. Much is now owned by conservation organisations and the wet meadows are famously rich in summer as breeding sites for a variety of wetland birds – particularly waders like Snipe and Black-tailed Godwits. The best natural sites are Wicken Fen in the east and Wood Walton in the west; they are mostly covered in vegetation and not open water. Wicken is particularly important as it was the first real nature reserve acquired by Charles Rothschild and donated to the National Trust – in 1899.

There are many other areas where wetlands of this type have been destroyed, or all but destroyed many years – often centuries – ago. For instance the Thames valley around and above London used to be marshland but, as it has been developed, and as the water table has sunk there is little evidence for this. The gravel areas, being close to a ballast-consuming conurbation, are now peppered with pits but these are not the sort of wetlands that have been lost. For many years these would have formed water meadows where grazing was used in the summer on ground which was too wet to cultivate in the winter and spring. Much of the Somerset Levels have been lost to peat digging; Martin Mere in

Lancashire is a small remnant of the local wetland and there are examples big and small in most counties. The marvellous Amberley Wild Brooks wetlands, where the River Arun is held up by the South Downs, is one of the few areas that has not changed much after the initial cutting of the drainage ditches – and it will not now be lost as it is a reserve area.

However it is not the large wetland areas which have been lost which have caused the general impoverishment of the environment all over the country. Individual small marshes – perhaps just an odd wet field here and there – have been drained and re-drained. Their loss has depleted breeding sites for such species as Redshank and Snipe to the extent that they are both now seriously rare breeding birds over most of lowland Britain. All is not lost as these sites can be made wet again by stopping the drains from working and ensuring that the water is held on site during the winter. Modern hydrological thinking is taking note of the very fast run-off of water leading to downstream floods and these new 'mini-wetlands' can prove to be useful in controlling the water. They will also develop into excellent botanical sites.

Upland marshes and water meadows

Many rough grazing areas seem to be continuous marshland. In some areas sedges grow amongst the grass and your stout walking boots fill with water after a few minutes. However there has been a tremendous amount of drainage work and grassland improvement in all areas. Ireland has been particularly astute in gaining money from the CAP but drainage has been a feature everywhere. The resultant improved grassland is often not just grazed. It may be cut for hay or, even earlier, for silage. Often it will no longer be wet enough for nesting waders, nor remain uncut or ungrazed long enough for species like Whinchats or Sedge Warblers to breed. The situation was very different earlier in the century when the agricultural industry was in severe depression and there was no pressure on agricultural land. The areas may look as they did in the past, but they are very different. The major give-away is the emerald green colour of the newly improved grassland and a much higher stocking rate of cattle and sheep. The latter also puts the breeding waders at hazard from trampling.

The lowland equivalent of these wetlands were the water meadows found all over the country along the river courses. These were important both as winter refuges for wildfowl and as rich breeding grounds for waders, particularly Lapwing, Snipe and Redshank and for Corncrakes. These birds also nested in drier hay fields but the hay was cut later in the wetter fields and so water meadows were particularly important. The water stored on the meadows in the winter was a very important flood prevention measure as it held back the flow preventing build up of dangerous amounts of water lower down the river catchment. One very important system, still in existence, is on the Shannon where the largest river in Ireland still has its 'callows' where the traditional management is still carried on and all these birds, including the Corncrake, breed successfully.

The fact of drainage and conversion to conventional farmland has made an enormous difference to the flood danger in many areas and the damaging inundations over the last few years are partly caused by these changes. The management of the water on the water meadows had been developed over many years with 'drowners' operating the sluices to make sure each area received its due portion. The flood also brought nutrients down from higher up which renewed the fertility of the fields. In many areas the water meadows had common ownership, with the rights to particular portions of the hay and to graze certain numbers of animals on the aftermath devolving to particular farms or cottages – and a large part to the lord of the manor.

Rivers, streams – and canals

From the broadest river, like the Severn, Thames or Shannon, to the tiniest stream, running waters are intrinsically very important for the birds they support and, particularly, for the bankside habitats they can afford. But much more than this they shape the landscape by funnelling the water that falls as rain back to the sea, carrying in them small fragments of the landscape – silt, sand, gravel, pebbles, stones and, in spate, rocks. So they are the means by which much of the country has been carved into the system of hills and valleys we see today. They have also provided the alluvial beds of sand and gravel that are used for ballast throughout the country – the pits are very important as sites for a very wide range of species (see below).

In lowland areas rather few of the rivers, or indeed the streams, are anything like natural. First of all the water flowing through the system is bound to be polluted with all sorts of substances. Our fresh waters are undoubtedly cleaner than earlier this century and, particularly in the first half of the nineteenth, when the Houses of Parliament had to be closed because of the stink from the Thames at Westminster about 150 years ago. In those days the sewage pollution, both human and animal, was a major problem but industrial activities were already producing absolutely horrendous chemical pollution as well. The running waters should contain a large biomass of animal life but many were absolutely sterile or reduced to stinking drains, homes to mats of bacteria and little else. The situation was gradually reversed and is now much better. How-

Tufted Duck and Pochard.

ever the distribution of Kingfishers, for example, is still mediated by industrial pollution with many individual waters, or whole areas, lacking in the birds because of poor water quality.

There is a further problem in the countryside from agricultural activities. These are really two-fold: waste pollution and run-off of fertiliser. Waste from intensive animal keeping enterprises is supposed to be carefully regulated and used to manure the ground, or to fuel power stations – however this does not always work. Where slurry enters small water courses it can devastate them and, until recently, the fines for such vandalism were often paltry. In fact as a business strategy flushing the slurry down the local river and paying the fine made selfish economic sense! Steps are being taken to ensure that the fines are much more realistic but accidents do still happen. There are also many small streams and ditches where pollution of this sort is all but continuous from animals and from silage clamps. This pollution is obvious both to eyes and nose but the run-off of fertiliser is more insidious. It even enters the ground water and there are many areas where the public water supply is dangerously high in nitrates. The increase in nutrients can cause changes in the ecology of the stream, at worst turning the bright and varied fauna and flora into flocculent mats of rotting material floating down a stream covered in a repulsive, metallic shimmering sheen.

However the ecological problems with the water quality are equalled, or excelled, by the physical changes which are routinely made to the bankside (riparian) habitats. Rivers and streams are very dynamic landscape features subject to flooding and drying out and generally with a wide sphere of influence often ten, twenty or a hundred times the width of the drainage channel. The channel itself is likely to alter radically over a short time and produce newly eroded banks, shingle bars, reed-beds and oxbow lakes. These may provide vertical faces for breeding Sand Martins and Kingfishers, open areas where Little Ringed Plovers and other waders can breed and the sort of vegetation favoured by warblers, Reed Buntings and a wide variety of other species. However the water itself is an important resource for man and it attracts development for homes and industry which, in turn, requires taming of the water-course. Often the river is continuously embanked with long length of concrete or other hard material used for confinement. The natural riparian habitat is effectively lost although birds like Sand Martin may nest down drainage holes and birds like wagtails and Dippers may be able to exploit holes under bridges and within brick work. There are some town centre Dipper nest sites in Wales and Ireland in completely built-up areas.

This loss of quality riparian habit is clearly a real problem in developed landscapes but it also occurs in many areas in the rural context. Here the problem has been caused by drainage and canalisation. For many years streams have been straightened and deepened in order to allow the water to flow away quicker and the surrounding areas to be tilled as arable land. Meanders and marshes have disappeared and the habitats have become simplified and impoverished. Indeed the works have produced their own problems in reducing the holding volume available to reduce floods further down stream. Such works are actually being reversed in some areas. At the BTO reserve near Thetford the meanders in the Little Ouse were restored some 40 years after the river was canalised: indeed the same person who drove the digger to restore the river worked on the original canalisation! The meanders provide excellent habitat for a variety of birds and the flooded fields are excellent for wintering birds and have had several waders breeding – including Redshank and Snipe.

Many people disregard canals as useful bird habitat but many are full of fish and, whilst one bank with the tow-path is generally hard and useless the other is often useful and 'soft'. There are many canals in southern Britain with stretches of linear reedbeds and thus linear colonies of Reed Warblers. Some basins are wide enough to harbour species like Tufted Ducks and almost all have Moorhens and Coots. Indeed canals have a very useful and special habitat at the locks where there is generally a small cascade of water. In lowland areas these are the only places where this regularly happens and these are the places where Grey Wagtails are likely to breed.

Ponds, pools, pits, lakes and reservoirs

It will be obvious that the most important part of most of the wet habitats, for us, is likely to be the sight of the water surface itself. The birds are looking at it in a very different way. The water surface may be used as a roosting site for the winter gulls, as a display arena for the Great Crested Grebes or Shovelers but for most birds it is primarily the upper boundary of the water body itself. The birds may be swimming on it but it is *within* the water that their food lives. Even the warblers living along the edge of the water are there because of the insects which breed in the water. Thus the quality of the water and the biomass of the different forms of life in it are vitally important to the birds. Every water body from the smallest pond to the largest reservoir and lake may be liable to pollution but they

are also characterised by their natural nutrient state. Upland lochs may be stunningly beautiful but very unproductive even though there is a tremendous flow of water passing through from peaty uplands – which contain nothing much more than the stain of the peat! Lowland lakes with much less of a throughput of water may be full of nutrients, not just agricultural runoff, and can support a very wide range and biomass of birds, other animals, and vegetation.

One just needs to compare the number of birds you will see on Llyn Brianne, a large upland reservoir on the Powys/Dyfed border, with those on the much smaller Tring Reservoir at Wilstone, on the Herts/Bucks border. The former is very much larger, with its surface at almost 1000 feet, and has a few dozen birds associated with it, and its surroundings, at any season. When I first knew the area, 35 years ago, the dam had not been built and there would have been many hundreds of pairs of 'land birds' living in the area. Wilstone Reservoir has been wet rather longer but is very much smaller and part of the lowland Aylesbury Plain. It has thousands of birds associated with it – ducks, geese, warblers, gulls, Swifts, grebes, buntings etc. – and would have had far fewer when it was unflooded.

The comparison of two man-made sites may seem inappropriate but, as cannot be emphasised too much, very many of the wetlands in the World are but transitory. Much open water silts up quickly and becomes woodland. If it does not, it is usually because the throughput of running water keeps it open. But in this case the open water is likely to change pretty catastrophically intermittently when there is flooding. Some upland lakes are likely to be more permanent but most suffer from fluctuating water levels as the outflow erodes the substrate holding the water in.

Many of the natural lochs in Scotland have raised beaches showing where the water level was much higher in the past.

The smallest open water habitats are the small ponds and pits which mainly provide drinking places for birds and are unlikely to act as homes for many, if any, true wetland birds. These used to be mainly farm ponds used for watering stock and the typical breeding species would be Moorhen and Mallard. Very many of these have been lost through filling in, pollution, neglect and development. However just as they have disappeared so we have seen the rise of the garden pond (see Human Sites). The role of the small water body – puddle, pond or stream – in the breeding dispersion of ordinary birds is little known but all sorts of observers have recorded very heavy use being made of these resources in the spring and summer.

Slightly bigger waters will usually have several wetland species associated with them if they are in productive lowland areas and have anything like natural banks. These waters are very often stocked with fish and used by anglers. This does not necessarily lead to conflict but all multiple use plans will involve compromise. For instance with fishing there will be the disturbance of the banks, management of the water-body itself and, possibly, perceived damage by the birds which predate the fish. On bigger waters various water sports are likely to take place. Even sailing dinghies will create a lot of disturbance but power boating, water skiing and jet skiing are very damaging. Water is a valuable resource for recreation and such pressures will increase, particularly near centres of population. For birds and nature conservation one of the most important considerations is coming to arrangements about the future of man-made water even before the digging starts. Far too often holes in the ground are dug to win gravel and ballast with the planning authorities insisting that they are returned to agricultural use – although the land is not of the best quality. Indeed the disposal of rubbish in landfill sites is worth many millions of pounds and subject to recent taxes – in turn partly used for environmental benefits. In certain circumstances the hole is more valuable than the material for which it was dug!

Water storage is a vital issue and in many parts of England and Wales the biggest blue areas (of fresh water) on the map are reservoirs. Some have relatively small lengths of embankment and long lengths of semi-natural riparian habitat but others are almost totally embanked. The countryside is seeing more and more farm reservoirs being constructed using plastic liners to keep water abstracted in winter for use for irrigation in the summer. These do not have natural bankside vegetation and can be of little use for breeding birds unless steps are taken to provide floating islands or other deliberately built habitats. Big lowland reservoirs are consistently excellent sites for birds and places like Rutland Water,

Abberton and Chew Valley reservoirs have a well earned international reputation. It is no surprise that all are characterised by areas of shallow reed fringed shores providing cover for the birds to breed.

These lowland sites are clearly rather special and also rather rare. If we go to the far North-west we come to areas which have lots of 'freshwater' blue on the map. In the Highlands and Islands of Scotland and in parts of Ireland – particularly the West – there are both large areas of water and areas where the land and water share the surface equally. For instance some parts of the Outer Hebrides are like this but these are not the famous shell-sand machair, with many breeding waders, but areas of wet, peaty land with rather few breeding birds. Even the very large areas of fresh water like Loch Ness and Loch Lomond in Scotland, Lake Windermere and others in the Lake District of England and the vast watery areas of Ireland – Loughs Erne, Mask, Corrib, Derg and the Lakes of Killarney – have rather few birds breeding on them although some of the Irish ones are crucial for Common Scoter.

The future of wetland sites as habitats for birds looks rather good. Many are already Nature Reserves and some are actually owned by conservation bodies. More pits are going to be dug for sand and gravel and local planners and residents are quite likely to welcome nature reserves and not want to have the holes filled with household waste! What is even more encouraging is the initiative being taken at places like Lakenheath, in Suffolk, where the RSPB is taking an area, drained for many years, and is returning it to wetland. Such initiatives could usefully be taken on both a large and a small scale all over the country.

Delany, S. 1993 'Introduced and escaped geese in Britain in summer 1991' *British Birds*: **86**, 591-599.
Kear, J. 1990 *Man and Wildfowl.* T & AD Poyser.
Owen, M., Atkinson-Willes, G.L. & Salmon, D.G. 1986 *Wildfowl in Great Britain.* Cambridge University Press.

Mallard.

MARINE AND COASTAL

The influence of the sea, on our climate, is absolutely crucial for every species breeding in Britain. But this is not what this chapter is about. Several million, the vast majority, of our seabirds are clumped in relatively few colonies on headlands and islands but there are some species which are more widely distributed. There are also many others, not typically thought of as being seabirds, that are heavily influenced by the sea.

Our sea-shore is very varied: the classic teeming cliff face colonies of seabirds does not represent the whole picture. Shingle shores and muddy estuaries are just as typical and have their characteristic breeding birds – but not in such numbers or so obvious. This chapter looks at the different habitats and the birds that they hold, the influences on those habitats – and the birds – from changes both natural and human. Loss of habitat, disturbance, development and pollution have had very bad effects but our major seabirds are doing well and are amongst the most important populations in Europe – and the World!

Shaping the high edges

The biggest influence the sea has had on Britain and Ireland is in carving us out as islands from each other and from the Continent. In fact the islanding of Britain was accomplished by the breakdown of a huge lake in the North Sea, held back by the ice. This eventually flowed into the Atlantic forming a huge cut through the chalk – the English Channel! And we are not talking about the dim and distant past but less than 10,000 years ago, after the last glaciation, and the huge lake was, of course, freshwater and not sea-water. Such a devastating, and almost instant, change to our geography has not happened since. The nearest example is probably the tidal wave caused by a long-run underwater landslip in Norway some four or five thousand years ago. This engulfed parts of the East Coast of Scotland with waves that seem to have reached 100 metres or more up the hills.

Everywhere the constant erosion, and accretion, caused by the sea, leads to changes which may be dramatic and obvious, or very gradual and imperceptible. For the generations of birds it is not the exact location of their breeding places but the habitat that is important. Research has shown that Guillemots are very likely to come back and occupy the same ledge from year to year. Indeed the female may lay her egg in the same place, to within a centimetre or two. The erosion, by a winter cliff fall, of the ledge would be inconvenient for the returning birds the next spring but they would be able to use newly exposed ledges or move to another part of the colony.

Prominent rocks, used by Gannets for breeding, like Grassholm off Pembrokeshire, Stac an Armin and Stac Lee at St Kilda and Ailsa Craig and Bass Rock off the West and East coasts of central Scotland, have probably changed very little since the seas became warm enough for the Gannets to breed in Britain, a few thousand generations ago. These solid and imposing sites are certainly not threatened with imminent disappearance by erosion. The geology of such sites is important. Gannets breed on isolated bits of land which need to be close to the preferred feeding areas. This is because the time budgets of the birds are crucial whilst they are feeding their young. However isolation from predators is also very important. Clearly we should think of man as being the worst predator but in the past the Gannets would have to be safe from wolves and bears as well as otters and foxes. An island with vertical sides, reasonably far off occupied land and with a flat top or broad ledges to breed on is ideal.

It is easy to apply the same constraints to where other seabird colonies are to be found. One can quickly see that there are likely to be mixed colonies as many of the birds are feeding in the same fish stocks. There are differences in the exact placing of the nests on the cliff – some are on small ledges, some in crevices, some in burrows, some on the grass at the cliff top. However they are generally clustered together as the colonies are often the only viable sites for the birds to breed and exploit the distant food resources. Some other species do not forage as far and birds like Black Guillemots and Shags, for instance, are found at larger colonies of other species in small numbers and also in scattered colonies away from the hot spots. The nocturnal petrels and shearwaters are always highly colonial and the islands of Skomer and Rum hold the two biggest Manx Shearwater colonies in the World each with over 100,000 breeding pairs. For these species and Puffins the important resource is often the burrow in the cliff top (or mountain top) turf and there have been erosion events which have caused big colonies to become extinct. For instance the peaty top to Grassholm was swarming with Puffins until about a hundred years ago when it eroded away and a colony which may have had 200,000 pairs now has only a few hundred – the birds now breed on the adjacent islands of Skokholm and Skomer.

These very hard, natural, coasts with rocks and cliffs are the first place one thinks about for nesting seabirds. Few of these areas are much modified by man – usually only a rugged lighthouse clinging to the headland – but this is not the case with much of the rest of the coastline. Areas of muddy saltmarsh have been claimed for agriculture, estuaries have been dredged and concreted over for use as ports and marinas and the open shore has been defended by seawalls and groynes. This means that the attraction they might have had for birds has been diminished. Very soft cliffs, subject to regular erosion on the South Coast, chalk, or along the East Coast, alluvial clays and silts, are seldom important for birds although some have been occupied by Fulmars and the softest sand areas may have Sand Martins. These areas along the East Coast are very important as donor sites for ballast which is washed up elsewhere and causes accretion. Where we have tried to defend the cliffs by hardening the landscape we have spent a lot of money and have often lost out. There are a large number of lost villages in the North Sea from the Holderness coast down to Dunwich in Suffolk: once a very important borough with 20 parishes and now reduced to about 60 houses.

Between a rock and a hard place

The high cliffs are not the only hard edge to the British Isles. For every spectacular breeding colony on cliffs, like those at Kebble on the Western end of Rathlin Island in Northern Ireland, there are great lengths of hard, rocky sea-shore with, at first sight, no birds breeding at all. These may be just as spectacular a vertical landscape as the colonised cliffs but many are simply rocky shores coming down to the sea with no particular

Rock Doves.

vertical component. These may have small colonies of some of the species found at 'real' colonies like Fulmars and gulls but they are often seemingly deserted by birds. This is not necessarily the case although, where there is a lot of disturbance and where small cliffs enter the sea sheer, it is true. In other areas the immediate shore may provide nesting and feeding grounds for such species as Red-breasted Merganser, Oystercatcher, Eider, Common Sandpiper and Shelduck.

However these are not the only beneficiaries. This sort of habitat is also the place to find Rock Pipits – indeed they never breed far from the sea and, as their name suggests, they like it rocky. It is also quite often an essential part of the coastal colonies of terns, gulls and skuas found in Northern areas. The birds themselves may not often be found feeding – or breeding – exactly on the interface between the sea and the land but many of the colonies are to be found on coastal areas or islands within a few hundred metres of the sea.

The open shore

The sea sloshes around a lot. Every month there is a sequence when the tides are higher than normal (the springs) and two or three times a year these high tides are worked up into a fury by stormy weather. Even more destructive is when the stormy weather and high tides are compounded by onshore winds and storm surges are created. One, almost 50 years ago, submerged a large part of East Anglia and Holland. Such events shape the landscape and provide the birds with renewed and new habitats. Not only do these events move material up and down the beach but they also impart an enormous lateral transfer and, with small particles, actually move fine sands and clays within the water column.

These natural processes are responsible for some of the very best bird watching spots in Britain – Minsmere in Suffolk and Cley in Norfolk for instance. However the processes have been looked upon as destructive of human resources and a huge amount of effort has gone into protecting the open shore from the sea. This has involved sea walls, the transporting of huge amounts of shingle from place to place and even the building of artificial reefs at sea to alter the deposition rate of the ballast being moved along the coast by the sea. The mobility of shingle, sand and stones can be seen all around the coast from the natural grading of the stones into different sizes along the remarkable Chesil Beach in Dorset to the 'fossilised' shingle ridges of Dungeness in Kent. Along the East Coast the strange landscape of Spurn Point is fed by the material washed from the Holderness coast and the Point would have become an island, disappeared and started to grow again long ago if it had not been defended. Further south the same material is building up at Gibraltar Point where the field

Ringed Plover and Little Ringed Plover.

station is in a building that was on the very edge of the sea when it was built earlier this century – now the sea is normally a mile further East!

The action of the sea at the land/sea interface depends crucially on sea level – and land level. The immediate interest is in sea level rises due to global warming (see later in this chapter and Climate and Weather). These will undoubtedly happen and have happened in the past. However one feature, going back a few thousand years, that many people do not realise has happened is the 'post-glacial bounce' that has caused the land to emerge from the sea. Basically the weight of glacial ice pushed the land down and it is gradually rising from the magma on which it floats. This is one of the reasons for the marvellous shore line habitats along the East Anglian coast.

The open shore habitats include the sand and shingle beaches and spits that are as crucial for some seabirds as the cliffs and headlands. Here Ringed Plovers and Little Terns breed just above (hopefully) the high water mark. Unfortunately in many cases they have human feet and the tyres of four wheel drive vehicles to contend with but many of the nesting sites are now in protected areas. This has increased the productivity of many colonies of Little Terns greatly by electric fences to keep out foxes, by supplementary feeding of predating Kestrels and by providing neat shelters for the chicks. The remotest shingle spits and ridges are a very important resource for breeding terns but ones near resorts are very often spoilt by holiday developments. In some cases military activities have protected these areas for many years – as at Spurn Point – or is continuing to protect them – as at Foulness in Essex.

The 'raw' open shingle spits and beaches, where the shore is accreting, quickly become covered with vegetation and provide very valuable habitats for a wide variety of small birds – particularly warblers and finches. But these sites do not just involve rough scrubby areas: they often are also wet and have reed-beds and other valuable habitats. There are also some areas where quite large areas of fresh (or brackish) water are naturally established behind the seawall. The aggregate that builds up is a very valuable asset if it is at all close to areas where there is civil engineering development. Winning these aggregates creates large holes which fill with water and may be used by many different breeding birds – including wildfowl, gulls and terns. In some areas there are lines of pits formed as the material was dug out to form the sea defences. The pits at Dungeness in Kent and Snettisham in Norfolk are good examples. Unlike inland pits dug closer to conurbations such holes are seldom used for landfill and, even if they are close, they may possibly be unsuitable because of the dangers of pollution.

These scenarios are often not allowed to develop because the shores are used for industrial developments, housing or as farmland. Remote coastal areas have been used in the past for nuclear power plants, as at Dungeness and Sizewell, or for conventional power stations and harbour developments, as at Shoreham in Sussex. Shingle beaches may not be comfortable for bare feet but long lengths of them are used as tourist bathing beaches – for instance most of the Sussex coast. Many areas of open shore are bordered by farmland which is seldom very productive. The raw shingle is converted to grazing land and many areas of shore are bordered by several hundred metres, to a few kilometres, of grazing marshes. In many areas these are limited by old cliff lines as in East Sussex, along the North Kent Coast and North Norfolk. There is one particular type of sandy habitat which is very special. This is the machair, consisting of calcareous sands which are typically very rich in flowers and, in some areas, have very dense populations of breeding waders. These sands have high densities of shell in them and can occur, in small patches, in many different areas. They can be very extensive along the West Coast of Scotland and Ireland – particularly the Outer Hebrides. Here the productivity of the machair has been enhanced over many generations by the use of stranded seaweed as manure by the crofters. The exceptional wader populations of the Western Isles are currently threatened by newly introduced or escaped mammalian predators – mink, polecat ferrets and, especially, hedgehogs.

Estuaries and river mouths

These areas are particularly important as habitats for birds in the winter when they are home, in Britain and Ireland, to about three million birds – particularly waders and wildfowl. In the summer, they are not so obviously important for birds. There are plenty of wildfowl, waders, gulls, terns and small birds breeding where they have the opportunity.

Indeed much of the area will actually be flooded at the spring tides and so cannot be used for nesting. Many of the Black-headed Gull colonies in such areas like the Solent, the Thames Estuary and the Wash are regularly engulfed by high tides. Even more important many of the estuaries, at least anywhere near areas where there are population centres, have been developed and have lost many of their natural habitats. For instance the Thames has much development and a very expensive barrage, to prevent flooding. However there are still large areas at the mouth of the Thames where the birds can breed. Indeed pollution levels are being controlled and commercial traffic is much reduced so the bird population is probably being encouraged. There are very good places on the North side of the estuary and such areas as the North Kent Marshes and the extensive reserves on Sheppey are excellent.

The characteristic of estuaries that is different from the open shore is the sediment. Inside the shelter of an estuary or river mouth the shallow water is likely to be moving very slowly for most of the tidal cycle and so small, and very small, particles come out of the water column and form mud and sand banks. These particles may have been brought in by the sea from elsewhere but most will have been brought down by the rivers (or blown off the land by the wind). These lead to a build-up of the land and, inevitably, its draining as agricultural or development land. The wild, unclaimed saltings are a very important habitat for breeding Black-headed Gulls and, particularly, Redshanks. The reclaimed land may be pretty uninteresting or can contain elements of wetland that make it very good for birds. One only has to think of such sites as the Solway, Morecambe Bay, the Dee, the Ribble, the Humber, Wash, Thames and Solent in Britain, and the Shannon Estuary, Cork Harbour, Dublin Bay and Wexford Harbour in Ireland, to realise the overall importance of this habitat.

Wet, wet, wet

Clearly the most important habitat for seabirds is the sea itself – although none has yet evolved a way of actually breeding at sea! Even as seafaring races most British and Irish people give very little thought to the seas round our islands as a habitat for birds. We only see the two dimensional interface that the sea makes with the air. The birds are dependant on a part of the three dimensions under it: what do they need and how and where do they find it? Most people have watched birds from the shore, from a headland or even from the deck of a ferry crossing the sea. Except for flying birds rounding a headland these are often pretty fruitless attempts at bird watching!

So the birds are obviously either spread out over the vastness of the ocean or clumped at places where they want to be and you cannot normally see them. Both answers have an element of truth to them. There is an awful lot of sea out there and the birds can be dispersed all over it. For instance Gannets can easily be picked with binoculars in good light two or three kilometres away. However they might forage up to 100 kilometres from the colony and so a coastal colony with 5,000 breeding pairs – dispersed over the possible feeding area – would have 5,000 adults (one of each pair would be on the nest) on 150,000 square kilometres! Or, to put it another way, you should on average be able to see one at any one time from a ship at sea. Other seabirds are much less visible and much less likely to be spotted. Indeed the fact that you do see clumps of seabirds at sea is an indication of what is happening below the surface – where the food is. We can see the changes in colour, lines of flotsam on the surface, changes in surface texture and even cetaceans feeding. These all give clues as to where the food may be and the birds use these clues – and much more.

Other species are very much smaller and there may be many more Guillemots breeding at a colony than Gannets but they are much less conspicuous. When one gets down to Storm Petrel size, (birds very like House Martins) the chances of seeing one, even if there are thousands and thousands around you, are pretty well nil. To be out at sea, with the land a long way off, and to discover small flocks of auks dotted all over the sea is an indication that you are sailing over dinner – for auks. You do not see the food and you do not see fishing boats exploiting it because the fish that the birds are eating will probably be the small sand-eel (*Ammodytes*) which, thank goodness, are not much caught by man in British Waters. However some populations in the North Sea are exploited for use as fish meal, fertiliser and, outrageously, as fuel for a power station in Denmark! I believe this stopped several years ago. These are the preferred food of many different species and they occur all round our coasts in populations that spawn in different areas of soft sea-bed. The fish are caught by the birds at various different levels in the sea and the recent poor performance of birds like Kittiwakes and Arctic Terns, which also feed on the same fish stocks, has puzzled biologists. Recently the suggestion has been made that global warming has caused the top few centimetres of the sea to become too warm for the fish so that the surface feeding birds cannot catch them! This would explain why these birds suddenly fail to find food in the middle of the breeding season when birds like Guillemots and Puffins are still coming back with plenty of food.

The three dimensional habitat of the sea is incredibly productive and there are still huge amounts of fish and crustaceans on which the birds can feed. The fact that many of the birds eat the same food during the breeding season is an indication that the separation of the species, by competition, may be because of their different feeding techniques or through competition for another scarce resource or, most likely, through competition for food at other times of the year. Many of the species that breed and feed together are to be found far from each other during the winter. Razorbills move further than Guillemots and Puffins shed their large, gaudy bills and take up a life of plankton eating in mid-ocean! Many of the birds positively benefit from the discards which the fishing industry provides and from the mess we make of the diverse ecology of the seas (see later).

The way that this food is dispersed at sea is very complicated. The fish stocks depend on the nutrients in the sea and they are, in turn, moved around by the ocean currents and the 'weather' in the water. Those clues that we get on the surface, the changes in colour and the line of flotsam, are an indication of what is happening below and the interfaces are often where the food can be found. Rocky reefs may be the places where some fish can be found but they are often where currents meet. The spawning grounds of populations of fish may indicate where the fish start off from each year but the larval forms of the fish, growing all the time, may drift around (in a fairly predictable way) with the water masses. Indeed the flightless moulting male auks, with their swimming chicks may simply

Choughs.

be swimming along on top of their food supply as they go across the North Sea from Shetland and then up the Norwegian coast. An understanding of the food supply is very important to understanding the bird populations and, unfortunately, rather little research is being done on these particular fish as they are not commercially valuable.

Pollution is not just oil

When one thinks of pollution at sea the immediate images are of oil and disasters etched on our memories – depending on how old we are – with names like *Torrey Canyon, Esso Bernicia, Sea Empress* and dozens more. Possibly even more distressing is the childhood memory of finding a bedraggled bird on the shore oneself. In fact pollution at sea is by a far wider variety of materials than oil and oil pollution varies depending on the type of oil involved, weather and season of the spill. Heavy, sticky oils may be dangerous to birds for weeks as they swill around being carried by currents and tides in the winter. Lighter oil, in warmer weather, may quickly evaporate or, in stormy weather, disperse. The awful effects of the oil pollution through the Second World War are largely unrecorded but tens of thousands of seabirds were regularly being oiled both as a result of routine cleaning of tanks and accidents during the 1950s and 1960s in British waters. The situation is much better now and the oiling rates of beached birds in most areas has been much reduced.

National and international legislation has had much to do with this as the majority of oiled birds are soiled with oil from tank cleaning (now illegal), leaks and even oil discharged (illegally) from land. Big wrecks hog the headlines and are responsible for huge amounts of oil being lost but these are getting rarer and less damaging; oil can be kept in better with double-hulled vessels. The recent awful oiling from the *Erika*, off North-west France, is a case in point. Several big oil companies had rejected the vessel for chartering purposes and general sea-worthiness standards are likely to be strengthened as a result of this incident. Such wrecks will become less damaging in the future as difficulties about responsibilities for decisions and the availability of powerful tugs is resolved. Of course the seas round us are now an oil producing area and there have been accidents with the newly extracted oils. Luckily most have not been large and those that have involve lighter oils which are not so dangerous. One aspect of oil pollution that is not realised by most people is that there are many time-bombs, large and small, on the seabed where vessels have sunk with oil on board. Sooner or later this is likely to be released as the wrecks break up. One or two very stormy winters in the last two decades saw an increase in the number of oiled birds washed up on lee shores and this could have been the cause. Several tens of thousands of birds may be involved and since it is the auks that are worst affected and the birds from particular colonies congregate in discrete areas the effects may be measurable at the worst affected colonies.

The damage done by an oiling incident is not necessarily related to the amount of oil released. A wreck involving 50,000 tonnes of light oil on a headland in very stormy weather might only involve a hundred or two birds. The very worst scenario with a narrow slick extending North to South down 500 kilometres of the North Sea in late July could actually affect several 100,000 auks. These would be the flightless moulting males and their chicks. Luckily such an event is pretty unlikely as the warm weather will evaporate the lighter fractions of the oil. Oiling affects the waterproofing and buoyancy of the oiled birds and is toxic to them as they try to preen it off. Cleaning of live rescued birds is expensive and very stressful to the birds – and rescuers – and is seldom successful as so many of the released birds do not survive. Prevention is by far the best solution to the problem. The use of dispersants to get rid of the oil has been very damaging in the past but new chemicals are not so bad and shores where the oil has ended up do recover after a few years.

Other forms of pollution range from the muds used for oil drilling, slag tipped into or by the sea from mines, sewage piped into the sea or dumped at sea, rubbish, agricultural chemicals and fertilisers, industrial effluent, wrecked or scuppered ships, explosives and even dumped chemical weapons and radio-active waste and discharges. If we have made it, then it has been dumped at sea! There is a huge volume of water out there but the sea cannot take everything thrown into it. The partly enclosed waters of estuaries are particularly susceptible, where the waters run off the land. These areas and inshore waters can have their balance of nutrients upset so that blooms of phyto-plankton produce toxins – red tides – that cause deaths of birds (and humans). Of course the extra nutrients from sewage can cause huge beds of shellfish to grow and this can be of great benefit to the birds – cleaning up the Thames has probably decreased the amount of food for the wildfowl and waders in the shallow waters and inter-tidal mud.

Chemical pollution of many types has hit in all sorts of ways. In Britain Peregrine Falcons which are feeding on oceanic seabirds build up dangerous doses of organic hydrocarbons in their bodies. These will have fallen as rain in the oceans and accumulated in the fat of the animal food chain eventually to be eaten by the seabirds and brought to the colonies. Mercury poisoning has proved to be a problem in many areas from the Mersey to Japan – and in the oceans (probably naturally). Polychlorinated biphenyls may have been responsible for seabird deaths in the Irish Sea over 30 years ago. It is possible that chemical pollution may be the cause of a general decline in the health of the fish stocks in the southern North Sea.

Plastic pollution may be very serious if the bits of plastic interfere with the digestion of the food by birds. For instance plastic film, as used in bags and packing, may be ingested and form an indigestible coating to the stomach. Another apparent problem allegedly was caused by small plastic items coming down the Rhine from an underwear factory. These small length of elastic material were nearly exactly zero buoyancy and 'swam' in the water column looking like small fish. Apparently they were exactly the size that Puffins prefer and the birds were found with these lengths of plastic in their stomachs undigested and occupying a part of the gut which should be providing nutrition. Similar problems have been reported from other oceans with zero buoyancy plastic.

A final rubbish problem, directly affecting breeding birds, is the use by many species that collect nest material of man-made rubbish in their nests. They would normally be using such items as seaweed and other plant material. These can provide excellent nests with no real chance of the nestling or the adults getting caught up in the nest. When the birds get hold of discarded, lost or damaged netting, or plastic string or other indestructible material there is a very serious problem with entanglement. In many Gannet colonies the nests are coloured blue, green or orange with netting designed to last for many years. At several colonies conservationists climb around the colonies during the non-breeding season removing this dangerous material.

And finally there is another form of pollution which directly affects the breeding birds. This is the infestation of breeding sites by predators brought in through man's activities. This has been very serious in many areas throughout the World. The most usual animal involved is the rat and the introduction has not been deliberate. Now some islands, like Ailsa Craig, have been the subject of successful eradication programmes and the species at risk may be able to breed much more successfully. Other animals involved include feral cats – often the fault of the lighthouse keepers – and native mammals as well as exotic or domesticated ones. For instance North American mink are now common in many areas and have had a very bad effect on gulls, terns, wildfowl and waders in, for instance, the Western Islands. Here they also have problems with feral ferrets and with hedgehogs which eat the eggs of ground nesting birds.

The effects of the fishing industry

Vast amounts of fish and shellfish are caught by the fishing industry with awful effects on the populations of many of the species being caught (possibly we should say most or all!). These activities can be very damaging in so many different ways it is almost impossible to start to document them. Heavy trawlers plough the bottom of the ocean to try to catch a few target species – rather like trying to catch butterflies in a meadow by driving a large bulldozer, blade down, across it whilst holding up a butterfly net. The damage done to the ecology of the soft ocean floor can be imagined. Dredging for shellfish can be pretty bad too where it is being done on an industrial scale. Many of the well known fisheries which used to be hand fished have suffered very badly recently.

We have regulations about the mesh size of nets, the take of fish and their sizes which are imposed nationally and internationally but they are ineffectual. Net-making technology can mean that the mesh of the trawl actually closes under catching conditions so that small fish are caught. Fish above quota are landed illegally or thrown back, almost certainly dead, into the sea. Small fish caught are likewise returned to the sea but are also probably dead – and if they are not they are probably going to be caught and eaten by the seabirds that gather round any trawler hauling her nets. This is a national and international scandal and requires the proper co-operation between the fishermen, the politicians and the fisheries biologists. One totally despairs when a politician (British from a previous administration) says something along the lines of – we have listened to fishermen and the scientists and will come to a compromise over quotas! These people employ the fisheries biologists to tell them what the fisheries can take in sustainable catch and then they decide on a compromise – outrageous!

So you may feel that the birds are in a terrible state. Wrong! The appalling mess we have made of the fish stocks in the sea has almost always been exceptionally good for the seabirds. The trick is that the birds have to be able to fly (we have no Great Auks left and penguins do not live in our waters). They are therefore quite small and exploit the small fish that were and are exploited by the larger fish whose populations we are messing up and therefore there are *more* not less of the very fish that the birds need. Populations are expanding as these small fish populations are increasing. An early result at Rathlin for Seabird 2000 is that the Guillemots have doubled and Razorbills increased by 70% in about 15 years. There are more and more Gannets in more and more colonies. This is likely to carry on as I fear we are not going to get fisheries policy right no matter how hard some people try.

The discards from trawlers have fuelled the increase in several species. Over the last 100 years Fulmars have gone from a species concentrated on St Kilda to one of the most widespread of cliff breeding seabirds. They are one of the species really doing well from discards and gulls and skuas have also done very well. Action is being taken to cut down on discards and this may have an effect on the populations of the birds that have become used to exploiting man in this way. The change of feeding strategy by Great Skuas on St Kilda and in Shetland may be partly driven by a lack of discards. Here there are recent estimates of annual kills by a relatively small population of this species of 40,000 and 200,000 seabirds respectively. This may drive down the populations of some species like the small petrels and Kittiwakes but this will be a natural change involving the interaction of native species.

Davidson, N.C. *et al.* 1991 *Nature conservation and estuaries in Great Britain.* NCC (now JNCC).
Davidson, N. & Rothwell, P. 1993 'Disturbance the Waterfowl on Estuaries' *Wader Study Group Bulletin 68 (special issue).*
Nelson, B. 1980 **Seabirds their biology and ecology.** Hamlyn.
Prater, A.J. 1981 *Estuary Birds of Britain and Ireland.* T & AD Poyser.
Lloyd, C., Tasker, M.L. & Partridge, K. 1991 *The Status of Seabirds in Britain and Ireland.* T & AD Poyser.

HUMAN SITES

The purist would, quite correctly, point out that the whole fabric of Britain and Ireland bears the imprint of man. Heaps of stones in the landscape might mark where prehistoric man had his huts or herded his cattle, where a cottage was abandoned during the potato famine or the Highland clearances – or just where some cowboy trucker fly-tipped some rubble. The lines across the hills might mark the pattern of Celtic fields dating back to the Bronze Age, where the Stevenson plough has prepared the moorland for forestry or the path taken by a four-wheel drive vehicle a few days ago. This chapter is not about these influences on our land but is particularly about what can be called the 'built environment'. To start with this means the obvious towns and cities, villages and hamlets but it also includes isolated buildings, industrial sites, quarries and sewage farms. Amongst the other activities included here are the transport and energy distribution systems.

It is easy to think of these habitats as totally negative. This is not the case and there are plenty of birds whose occupation of buildings for nesting is the rule and not the exception – for instance Swifts, Swallows and House Martins amongst the common birds and Black Redstarts and Barn Owls amongst the rarer. The important point to take on board is that the birds look at the available habitat from the point of view of what it offers them and what they can exploit. Looked at this way it is clear that we lost a really good opportunity when building such massive structures as the Dungeness nuclear power stations by the sea *without* ledges. There might have been a whole new seabird colony in Kent! And we should remember that the birds do not have a sense of aesthetics when it comes to choosing a home – run down areas of decrepit housing may be far better for them than the equivalent area which has been well looked after and maintained. For instance I and my friends in *Concern for Swifts* (leaflet *Concern for Swifts* available from the RSPB or BTO or phone 01353 740540) feel that one of the factors driving a disastrous decline in this species, in many areas, is simply better roof maintenance. This is a great shame as it is very easy to have an excellent modern roof and still allow the Swifts entry.

Black Redstart.

Human habitation

For the birds the bleakest of housing estates or city centres must be rather like a limestone cliff landscape without the botany. As we might expect there are isolated cliff nesting species – like Kestrels and, probably sooner rather than later, Peregrines. However the extra attribute that the human sites have is rubbish – even if they do not have any natural vegetation or planted parks and gardens. This rubbish may look very unsightly and unpleasant but it can include all sorts of food – restaurant debris, discarded chips, dropped sandwiches and bits of hamburger. These can be, and are, exploited by such species as Feral Pigeons, House Sparrows, Starlings, Jackdaws, gulls and even (in the past and possibly in the future) Red Kites! These birds not only use the rubbish as food but often they find useful items of nest material – there are even records of complete nests made out of discarded electrical wiring! In fact built-up areas usually have trees and gardens and this means that the habitat resembles wooded gorges and canyons. So extra species can make use of the habitat – provided they can learn to ignore the activities of the people living and working in the area, the sound and motion of vehicles and to keep out of the way of cats! Birds able to do this in many areas of Britain include Blackbird, Blue Tit, Great Tit, Wren and Robin. These are largely species of the woodland edge and the trees and shrubs that are deliberately grown or grow up unasked are very important features. In most British urban areas there are likely to be excellent areas of public parks which may be rather regimented but, nonetheless, offer such species as Dunnock and Chaffinch a complete home and species like Carrion Crow and Rook tall trees in which to nest. There are many bigger parks with lakes and islands which offer wildfowl the opportunity of nesting. London has a very good example with Grey Herons breeding on the willows growing on the islands of the lake in Regents Park!

In all but the very centre of the biggest cities there are many other species. They now include the recovering Sparrowhawk, which was nesting in central Bristol and Edinburgh fifteen years ago, and such migrants as Swift, breeding in buildings, and House Martin, breeding on them. These birds have been able to move to inner areas following the cleaning up of urban air and the huge drop in the use of coal for domestic fuel. Both species rely on the architecture of the buildings and the martins are only able to nest if there is a supply of suitable mud within 200 metres of the nest site – as at Rotherhithe where the Thames presumably provides the mud. We could, of course, give the birds a helping hand by erecting nest boxes. Birds such as Pied Wagtail, Song Thrush, Collared Dove, Magpie, Mallard and Tawny Owl can be found nesting on buildings in these areas. In some towns, and not only those close to the sea, Herring and Lesser Black-backed Gulls may nest on the roofs. These are still in areas which, from the air, are mostly grey and red from building rather than green from trees and gardens.

Small towns and country villages have an even greater chance of hosting different species of breeding bird. Grey Wagtails and Dippers nest near streams and rivers on structures – often bridges and culverts – in upland areas in Britain and also on many quite slow, lowland rivers in Ireland. Treecreepers and Nuthatches are common garden birds in wooded areas but they will be dealt with later. Perhaps the place that we think of as the 'type nest site' for the Spotted Flycatcher is the space where a brick is missing from a rectory garden wall. As one gets into the more rural areas so the association with House Martins and Swifts also embraces Swallows and this is probably the best place for House Sparrows and, in a few favoured areas, Tree Sparrows to be nesting in buildings.

Isolated farms and farm buildings are even better for birds. All the species mentioned so far can be present but in better numbers. Farms in the West of Ireland and in parts of Scotland can still regularly boast ten or more pairs of Swallows – something that would be very unusual in southern England where there is now rather little livestock farming. Barn Owls regularly use barns for breeding and some are specially designed to allow them in – for instance in the Borders around Dumfries most barns have an owl hole. Other species which use this sort of building include Stock Dove and even Shelduck. Piles of bales, stored for several months and left over the summer, are regularly used as nest sites by Barn and Little Owls.

Industrial developments

The BTO has run several challenges to seek out good industrial sites for birds and an astonishing range of different sorts of places have been entered. Quite often extensive industrial works have actually been built on valuable sites for nature conservation which would, in their natural state, now be protected. Often not all the interest has been lost as many sites need to be 'insulated' from other people and some are really very well protected and hardly managed at all – for instance areas where explosives are being manufactured. However these sites are not really interesting from the point of view of their built environment but as a result of the rest of the habitat. The same can be said of military training areas – in many cases the next best thing to being a nature reserve, as the training itself seldom causes as much damage as the use of the area for other purposes would involve.

The actual buildings at industrial sites may afford all sorts of opportunities for birds to nest. Ledges on buildings are often colonised by everything from the Feral Pigeon to the arch pigeon-eater, the Peregrine! Indeed the Peregrine looks poised to spread across lowland Britain as it has made up the losses suffered from pesticides and benefits from protection. The idea that they should nest on buildings is not new and Salisbury cathedral had a pair centuries ago. Now cooling towers at power stations, tall factories, flats and even pylons are being colonised provided that there are nesting sites provided – and they are being put up in more and more cases! Pipework in many plants provides excellent nest sites for species like Starlings, Jackdaws and Pied Wagtails that like to nest in holes and crevices. Shingle nesting birds, like Oystercatchers, may even nest on roofs – along with gulls – and they and other species may use car and lorry parks and other shingle and gravel areas to nest. There are coastal sites where Kittiwakes nest on dock walls, old castles and warehouses – I have even seen a nest on a lamp post in front of the colony!

Extensive industrial sites

Industrial sites include rubbish tips, sewage treatment works, quarries and even reservoirs. These all have particular attributes which appeal to different sorts of birds. What is a nasty looking mess, to the human eye, of a rubbish tip may well provide the local Kestrel and Tawny Owl with an à la carte selection of small mammals. Indeed on a trip to Finland I was told that the territories of their Eagle Owls are often centred on the local rubbish tips! Sewage treatment works often have running water and may provide the only suitable sites for breeding Grey Wagtails in lowland areas and they are often the homes to several pairs of Pied Wagtails. Modern sewage treatment works are very different from the old-fashioned ones which had extensive areas of marshland – 'tanks' – where the surplus water was stored when there was heavy rain. The really old style farms were very good indeed as the idea was that the effluent would be allowed to settle out on fields which were cropped every three or four years. The glories of Wisbech, Peterborough and Cambridge sewage farms will be etched on the minds of ageing birders who knew them. Breeding Snipe, Redshank, Lapwing and Little Ringed Plover were just the start. Birds of rank vegetation like Grasshopper Warbler and Whinchat would sometimes turn up and, of course, there were Moorhens and other wetland birds – as well as feeding Grey Herons.

Quarries are a particular love of mine. I spent several months of nights camping out in gravel diggings and sand pits – usually under the stars – in the summers of the 1960s when we were ringing Sand Martins. In the Home Counties probably 99% of these birds, then more common than they are now, were nesting in pits and most of the rest nested in drainage pipes rather than really natural sites. We found everything from multinational sites where there were investments of many millions of pounds, through family firms and down to 'parish pits' where local inhabitants were allowed, by right of abode, to come and dig their own sand. The variety of wonderful habitats these represented can be imagined – everything from wide expanses of water, vast reed beds, shingle banks, acres of wet scrub, woodland and so on. The breeding birds that used the sites 'by mistake', as the owners were seldom at all interested in the natural history of their areas, were excellent. We had Kingfishers, Little Owls and Tree Sparrows breeding in the Sand Martin holes. The local birds included Red-backed Shrike, Nightjar and the very rare Dartford Warbler as well as many birds that would not have had a chance of breeding in the farmland had it not been dug to win the minerals.

Many of these sites, dug more than 35 years ago, no longer exist as working extraction pits. Some are built on or have been filled in, often with rubbish, and returned to boring farmland. But many are still in existence as nature reserves, fishing lakes, sailing or power boating sites and even as water features in new towns. We should be very pleased to accept such areas with open arms but we should not *expect* that they will become nature reserves simply because the working site is colonised by good birds or rare plants. The planning permission for the site may have been granted provided it was reclaimed later – indeed in some cases the value to the owner of the disposal of the rubbish in the landfill may be greater than the ballast extracted. As conservationists we should be looking out for planning applications and getting the eventual use of the site for nature conservation purposes written into the original permission.

This may be surprisingly easy as it may ease the planning procedure for the owner and the gift of the site to the local Wildlife Trust, at the end of the sand or gravel working, may result. If this is the case the landscaping of the pit may be possible, as the ballast is won, to make it better for birds. For instance it is quite usual to have 5% of the area of flooded workings given over to islands.

Whilst sand and gravel diggings are very likely to be good natural history sites, open cast mining for coal or iron-stone and the pits for china clay are generally pretty dire. Clay pits for bricks, on a small scale, are often quite good and some of the bigger ones do grow reasonable vegetation or develop into good wetlands but they seem to take longer to vegetate. Stone quarries are likely not to become such good sites quickly as the hard substrate takes a long time to grow a cover of plants but small pits, worked for many years, become very good and, like the small gravel pits, often provide the only semi-natural habitat, apart from woodlands, within the mosaic of farmland. In some areas the actual physical provision of properly vertical cliffs, in an otherwise eroded landscape without safe ledges, is very important for breeding birds like Ravens and Peregrines that like such sites. Similarly, colonies of Jackdaws may form in the crevices and crannies made available by the mining.

Transport – vehicles and power

The covering of Britain with tarmac took place largely during the last century – not just through roads and car-parks but also through airfields. It has been more of a problem in areas where there are large numbers of people and is not a problem in any of the upland areas or in most of Ireland. The criss-crossing of most of England by railways happened rather earlier and now many of them have been closed and goods yards reduced. Ship-borne cargoes are now not using the traditional ports but new container terminals and so the old docks are being re-developed whilst some very good natural sites are being destroyed. Huge numbers of wires are strung across the countryside from poles and pylons to supply electricity and telephone communication. Their striding lines are not only intrusive but many of them have to have the vegetation along the lines regularly cut. At intervals there are aseptic and ugly switching stations and transformers. In recent years mobile phone aerials have sprouted all over the place without any central planning of their location, or until recently, any co-ordination between the companies concerned. The provision of pipelines for gas, water and oil provides temporary disruption but often the scars heal completely – the same cannot be said of the pumping stations and processing plants.

Roads are a pretty unmitigated problem for our birds. A few are able to feed on the dead wildlife mown down by the traffic – from small birds on insects to scavengers on rabbits and even larger animals. But these birds are quite likely to become a part of the crop of carrion themselves. In fact it is much, much worse than this. The birds have lost habitat to the tarmac and they will lose their lives, or the lives of their offspring, to the scavengers that are able to survive the winter on the corpses thoughtfully provided by our cars! Direct bird deaths from RTAs (road traffic accidents) probably amount to 30,000,000 to 60,000,000 a year in Britain and Ireland. Birds like House Sparrow and Blackbird are prime losers but there are huge losses of such birds as Barn Owls. About 20% of the nestlings we ringed here in Norfolk one summer were reported dead on roads within three months of ringing and bird hospitals record many more right wing breaks than left. A Barn Owl taking off from a perch beside the road will be lower across the near carriageway as it gathers speed and so is likely to be hit from the right.

The menace of road deaths as a result of moving traffic has increased greatly with the increasing speed of traffic. At about 45 m.p.h. the approaching car is travelling faster than an avian predator and birds do not seem to be able to react in time. However research has shown that birds are likely to shun really busy roads as the noise of the traffic interferes with their communication and so the real menace is not the motorways and busy trunk routes. Rather it is the country lanes and A roads and the suburban streets where the carnage is at its worst. There are arguments for suggesting that roads with hedges on both sides are the worst of all as the birds will fly from one side to another – and get wiped out. This is the cause of death for many species like Whitethroat which are likely to have territories spanning the road. Another really lethal situation, for walking birds, is where there are two ponds or marshy areas each side of the road – Moorhens are the victim in this case. Individual motorists can do their bit by flashing lights at birds on the road – this makes them look at the approaching vehicle to appreciate its speed since they are then using binocular vision. One lorry driving correspondent wrote to me of over 150 dead Barn Owls he had seen on the road in the course of three year's driving. This is about 2% of the breeding adult population at present! Peripheral help is given to birds by the presence of nesting sites in culverts, under bridges and on viaducts. There was a famous Red-backed Shrike's nest actually on a huge roundabout near Winchester but this was a real exception.

In contrast to road, the declining railway system is almost a benefit. Rather few birds are killed by speeding trains, the trains are well separated in time so that noise is not a problem and railway habitat is so much better than that provided by roads. Quite large areas of embankments and cuttings, as well as abandoned marshalling yards, are to be found throughout the system. Small birds nest along them and often quite rich areas of trees and shrubs are allowed to grow. In the Home Counties several pairs of Nightingales were reported nesting along railway lines in the BTO 1999 survey. There is even a Grey Heron colony on the Pymore crossing of the Ouse Washes. The actual ballast of the track, even when in spasmodic use, may be chosen for nesting by waders – for instance Little Ringed Plover and Oystercatcher. Airfields and their buildings are not very good for birds as they are not good flying companions for the aircraft. Every modern airfield is actively managed to make it an unwelcoming place for birds. There are, however, many smaller airfields with nesting waders whose adults are fully apprised of air traffic control procedures! In addition there have been large numbers of military airfields, surplus to requirements, which have been the long-term homes to many open country birds – including such rarities as Stone Curlew.

Power lines and telephone wires, and their pylons and poles, are a hazard to migrating birds but are normally well known by the local inhabitants. There are some species that may suffer from deaths through local collisions – these include Mute Swans and large grouse. A very significant number of Mute Swans die through wire collisions but are not at risk during the period when they are raising young – the adults undergo a complete moult at this time. The supports for the wires can be used for nesting – without modification. Carrion Crows and, in a few cases, Rooks, may use them. Many other species can be enticed by nest boxes or nesting platforms. Indeed it is possible that pylons will allow the spread of Peregrine Falcons across lowland Britain and Ireland.

Gardens

Of all human habitats, apart from nature reserves, gardens are the most likely to be managed with birds as a partial priority. The huge extent of gardens, if you consider them together, and their wide dispersion makes them very important. After all the first Collared Doves to colonise Britain were in gardens and their initial spread seems to have been associated with backyard chicken keeping. The vast gardens associated with stately homes and municipal parks seem to be rather a special case but they are really not much different, for the birds, than a clumped amalgamation of several adjacent smaller plots. Structurally the average gardened area is usually more or less open woodland with stretches of low vegetation – the borders, beds and grass. Most gardens are subject to human disturbance – clearly this is sometimes prolonged and rather devastating if it is a small garden and there are large numbers of children! Disturbance by cats and dogs can also reduce the value of the garden for birds but the basic determinant of the value of the garden is the habitat – and this does not mean only the garden itself but the adjacent gardens too. Most birds will have breeding territories which extend over several normal sized gardens. This is one of the reasons that people are often disappointed by the persistent cold shoulder that their nest boxes get from the local tits. The birds may be well suited by a site – nest box or otherwise – in an adjacent garden!

The rise and rise of the garden as an important habitat for birds is best illustrated by the summary below of the use made in winter of gardens measured by the Garden Bird Feeding Survey run by the British Trust for Ornithology and by the equivalent in Ireland.

Species	1970s records		Recent records		Ireland 1997/98	
	%	Rank	%	Rank	%	Rank
Blackbird	99.3%	1	99.5%	1	99%	1
Blue Tit	99.1%	2	99.5%	2	99%	2
Robin	98.9%	3	99.2%	3	98%	3
Great Tit	93.0%	7	97.1%	4	92%	6
Chaffinch	92.1%	8	96.4%	5	96%	4
Greenfinch	91.5%	9	96.3%	6	93%	5
Dunnock	95.2%	6	95.7%	7	78%	10
House Sparrow	96.6%	4	90.2%	8	78%	9
Starling	96.4%	5	90.1%	9	70%	14
Collared Dove	60.1%	12	87.1%	10	52%	19
Coal Tit	69.8%	11	83.8%	11	88%	8
Magpie	29.4%	17	71.7%	12	91%	7
Song Thrush	88.4%	10	61.7%	13	73%	12
Woodpigeon	19.3%	20	53.4%	14	60%	16
Wren	33.6%	15	50.8%	15	77%	11
Sparrowhawk	9.7%	28	48.9%	=16	33%	24
Jackdaw	32.4%	16	48.9%	=16	72%	13
Siskin	7.3%	29	48.8%	18	55%	18
Long-tailed Tit	11.4%	25	48.2%	19	25%	25
Goldfinch	3.1%	33	42.3%	20	37%	21
Pied Wagtail	44.9%	13	40.5%	=21	47%	20
Great Spotted Woodpecker	19.6%	20	40.5%	=21	None in Ireland!	

Starlings.

Species	1970s records		Recent records		Ireland 1997/98	
	%	Rank	%	Rank	%	Rank
Carrion/Hooded Crow	15.6%	23	33.6%	23	46%	21
Rook	21.4%	18	23.8%	24	65%	15
Nuthatch	17.9%	22	23.4%	25	None in Ireland!	
Mistle Thrush	34.6%	14	23.2%	26	35%	23
Jay	10.4%	27	19.7%	27	2%	38
Pheasant	5.9%	32	19.2%	28	5%	31
Feral Pigeon	6.1%	31	15.0%	29	9%	28
Bullfinch	21.0%	19	14.5%	30	58%	17
Yellowhammer	6.4%	30	12.6%	31	4%	34
Reed Bunting	13.5%	24	10.8%	32	6%	30
Tree Sparrow	10.8%	26	10.5%	33	3%	35

Figures are from the Bird Table Book for 723 annual records from feeding stations in the 1970s and more recently for about a thousand from GBFS master sheets for the last four winters of the 1990s. The Irish data are from over 527 gardens in the republic counted in December 1997 and January and February 1998. Marsh and Willow Tit are omitted because of the difficulty of identification and species that are largely winter visitors – like Black-headed Gull, Fieldfare, Redwing, Goldcrest and Brambling are omitted together with species with less than 10% 'Recent records'. The figure given is the percentage of gardens that had at least one record of the species during the six months of record keeping over each winter. The Redpoll (21%) comes into the Irish list at well over 10% but only featured in the British lists at less than 1% in the

Robin.

earlier period and at 2.2% more recently.

Obviously there are clear similarities between the three lists and the top three species are the same. The next three in the current 'pecking order' in Britain have increased in percentage terms – all are now absent from less than half the percentage of gardens where they used to be missing. Dunnocks, not noted for bird table use, have marked time, but the next two – House Sparrow and Starling – are real losers. These are poorly represented in Ireland. The top 12 in the 1970s are rounded off by two species which have increased markedly – Collared Dove and Coal Tit – and one in severe difficulty – the Song Thrush, now found in three times less gardens than it used to be. Almost every species is now found in a greater percentage of gardens than it was 30 years ago. There are some species in Ireland which are found in a much higher proportion of gardens than in Britain – these include Magpie, Wren, Jackdaw, Rook, Bullfinch and Redpoll. But there are others with a much lower proportion of gardens occupied – Dunnock, House Sparrow, Starling, Collared Dove, Sparrowhawk, Long-tailed Tit, Jay, Pheasant, Yellowhammer and Tree Sparrow – not to mention the two species absent from Ireland. In Britain, where we can compare the prevalence of the species over an interval of 30 years, taking the species which used to be in less than half the gardens and calculating the ratio between the percentage now and then we can compare them:

Species	Ratio 1990/1970 gardens	Current percentage
Goldfinch	13.65	42.3%
Siskin	6.68	48.8%
Sparrowhawk	5.04	48.9%
Long-tailed Tit	4.23	48.2%
Pheasant	3.25	19.2%
Woodpigeon	2.77	53.4%
Feral Pigeon	2.46	15.0%
Magpie	2.44	71.7%
Carrion (Hooded) Crow	2.15	33.6%
Great Spotted Woodpecker	2.07	40.5%
Yellowhammer	1.97	12.6%
Jay	1.89	19.7%
Wren	1.51	50.8%
Jackdaw	1.51	48.9%
Nuthatch	1.31	23.4%
Tree Sparrow	0.97	10.5%
Pied Wagtail	0.90	40.5%
Reed Bunting	0.80	10.8%
Bullfinch	0.69	14.5%
Mistle Thrush	0.67	23.2%

Many of these species have increased in gardens at the same time as their overall population level has increased – Sparrowhawk and Magpie for example, or declined as it has declined – Mistle Thrush and Bullfinch. However there are species which have done particularly well as the sort of foods that they like have been offered in gardens – Goldfinch and Siskin are examples. Perhaps the most interesting ones appear at the bottom of the list – Yellowhammer, Tree Sparrow, Reed Bunting and Bullfinch. These are seed-eating birds whose overall populations fell by 60%, 87%, 64% and 62% respectively over the 25 years from 1972 to 1996. Allowing for the population changes British gardens got more popular for Yellowhammers by a factor of nearly 5, for Tree Sparrows by about 7, for Reed Buntings by about 2.2 and for Bullfinch by 1.8.

This clearly shows that many people are getting it right in their gardens – if they are trying to increase its attractiveness to birds. They are largely doing this by offering food, and offering a more varied selection – and so attracting a wider variety of birds. Thirty years ago most keen bird feeders would be offering peanuts, corn tailings (or other general seeds) and fat. Now sophisticated suet products, live foods like mealworms, a wide variety of seed mixes, niger and, particularly black sunflower seeds are in wide use. There is also the very important change from sporadic feeding in bad weather during the winter to round-the-year feeding. This does encourage breeding birds by providing food for the adults whilst they can feed their young all the 'natural' foods they can find. This is particularly important since our gardens are not ideal places for many of our native birds to breed. We tend to object to insects breeding on our flowers and, in any case, the varieties and species we grow are often highly modified from the original wild plant and may originate from thousands of miles away.

One example is the dilemma facing Blue and Great Tits breeding in gardens. Here there may be thoughtfully provided nest boxes and lovingly tended trees. However the trees are unlikely to be the forest species – beech, oak and ash – that normally provide the parents with a super-abundance of caterpillars for feeding the parents and their chicks. Provision of peanuts (behind mesh) and seed allows the parents to feed themselves quickly and spend more time finding insect larvae to feed their chicks. A consequence of this is that productivity of garden nests is much lower at 4, 5 or 6 chicks than woodland nests which may be as

Coal and Great Tit.

Blue Tits.

much as 10 per nest. The immediate thought is that we should not be feeding birds to attract them to gardens where they do badly but should be scaring them away to nest in natural woodland! However this is wrong. Productivity is lower but the full grown birds survive very much better in gardens – being provided with food and so live longer than their country cousins. This seems to be the case for a variety of species which have been studied in detail. Once the suburban bird has learnt about cats, cars and windows it has a fairly cushy life.

The downside of gardens is that many are a mosaic of very small and intensively used individual plots with little structure to them. This means that the open nesting birds have little opportunity of nesting in secure places unlikely to be found by the prying cat or searching corvids – mainly Magpies. The lot of the suburban nesting bird would be made much better by the planting of more prickly shrubs, evergreen bushes and thick creepers. These would provide multiple nest sites, excellent roosting possibilities and, in some cases, extra food supplies. If asked to name the top choices I would go for ivy and honeysuckle for the climbers and dense varieties of berberis, cotoneaster and pyracantha for shrubs.

House Sparrow.

Prospects for the future

More and more people are feeding the birds in their gardens and taking notice of wildlife in general – this must be a good thing for the birds. Not only are people taking more care in their gardens but the principles of conservation, as applied to planning, farming, building – general behaviour – are not only becoming important to more and more people but they are being enshrined in law too. This does not mean that everything is rosy. Appallingly bad decisions are still taken – like the barrage at Cardiff Bay – and the law needs to be tightened up. A totally unacceptable number of Sites of Special Scientific Interest in Britain and Natural Heritage Areas in Ireland, together with other designated sites, are damaged each year. The principle that these are national assets needs to be accepted by all parties. A very important principle, that would defuse many confrontations, is that green field sites which are to be used for development should be just that! Fields mean farmland, farmland, in most cases, means that there is little of conservation interest left and so, given adequate compensation to the farmer, the ecological loss is limited. The contribution to the overall wildlife interest of the countryside of intensively farmed fields is minimal in many areas compared to the hedges surrounding them. Of course development does mean that it is unlikely that they would ever be returned to a better state. On the other hand the unproductive, in agricultural terms, parts of the countryside are generally very much more valuable for conservation. Roads through ancient woodland or across marshes are a different matter, and the protesters in these cases have a very much stronger position.

I suspect that we will see much greater provision of nesting sites within the built environment as a means of 'environmental mitigation'. For instance Cellnet now routinely erect a Kestrel box and two tit boxes on each new mobile phone aerial. Redevelopers of existing sites where House Martins or Swifts breed are attempting to ensure the birds are able to come back and Norwich City Council is ensuring that pantile roofs that they maintain have Swift boxes inserted. Perhaps we will see Kestrel and Peregrine nest sites built into high buildings and flat roofs shingled for waders to nest. Indeed future supermarkets might be built with Swift nest sites ready for occupation so that the icon of a British summer will gradually change from the parties of swifts screaming around the ancient church tower to gatherings round Tesco or Sainsbury – just like the human inhabitants!

Cannon, A. 1998 *Garden BirdWatch Handbook*. BTO.

Glue, D. 1982 *The Garden Bird Book*. Macmillan.

Hoskins, W.G. 1977 *The Making of the English Landscape*. Book Club Associates.

Mead, C. 1997 ' Pathetic Bundles of Feathers – Birds and Roads' *British Wildlife*: 8, 229-232

Nicholson, E.M. 1995 *Bird-Watching in London*. London Natural History Society.

Oliver, P.J. 1997 ' The breeding birds of Inner London, 1966-94' *British Birds*: 90, 211-225.

Reijnen, M.J.S.M., Veenbaasm G. & Foppen, R.P.B. 1995 *Predicting the effects of motorway traffic on breeding bird populations*. DWW, Delft, Netherlands.

CLIMATE AND WEATHER

Few people would dispute the assertion that in Britain we are obsessed by the weather. Now we have also become obsessed by the climate as well; global warming has, in my view, long been proved to be real. The El Nino event of the late 1990s was a feature of numerous news stories. The huge effects of climate and weather on our birds cannot be over-emphasised. The changes that we are seeing worry us as we seem to be responsible: our activities, running out of control, are poorly understood. However climate change is normal; it will mean that our avifauna changes – but these, or other, changes might have been coming anyway. To understand the *State of the Nations' Birds* we need to understand the influences of climate and weather.

My reaction may seem to be rather less than the panic you might expect; but the simple historic fact is that the last glaciation did not end until some 10,000 years ago. The pleasant woodlands of the Peak District were under permanent ice. The monumental landscape of the Highlands was being ground out by immense glaciers. Southern England was attached to mainland Europe and bits of East Anglia were integral parts of the Rhine delta. Our ancestors were hunting an exotic range of species including animals now long extinct. The warming that has been going on since then has not been steady and there have been periods when the weather has been much colder. The succession of winters, about 250 years ago, when ice fairs were held on the Thames came to be known as the 'Little Ice Age'. These long term changes are becoming better understood through the study of tree ring records and the ice record from the Greenland (and other) ice caps. We should remember that thermometers are quite modern inventions and that good weather records only go back a couple of centuries or so.

The possible consequences of the climate change which we are experiencing will be dealt with at the end of this chapter. For the moment it is the more immediate effects of the weather on bird populations that deserve our consideration. Incidentally, 'climate' is what the weather is generally doing over several decades – a sort of rolling mean, as it were. 'Weather' is what changes from year to year and season to season and is – usually – the real determining factor that drives the year-on-year changes in bird populations. In many cases the annual changes are statistically significant and the changes in breeding populations for many species may be reduced by a third, even two thirds, as a result of a very cold winter.

Winter

The winter weather can create direct mortality for many, if not all, species and may so debilitate the birds that do survive, that they breed late – or not at all – in the subsequent breeding season. Of course there may be the other effect: that the surviving birds breed very well in the absence of competition from the birds that have died! The cold can kill birds outright but most can survive if they can find and eat sufficient food. This may seem easy as we see thrushes like Redwings, Fieldfares and Blackbirds tucking in to windfall apples in the snow. However these birds would be feeding on soil invertebrates with far higher nutritional values if the weather was not freezing and fruit is a poor second best. They try to lay down fat as insurance during the winter so that they can, to a large extent, live off their reserves and only have to top this up with the huge volume of fruit they eat. As one can see from the loose droppings this is mostly water. Of course most birds are feeding on a diminishing resource during the winter. Plants are not producing extra seeds and the insects are not growing. In woodland the tits reduce the resources as they rush round feeding in flocks. There is one place where this does not happen – gardens! Here regular feeding provides a food resource that the birds can rely on to be there – as long as we keep feeding. There is good evidence that tit survival is very good in gardens but their breeding productivity is poor – gardens generally lack the big native trees that produce the flush of caterpillars needed for feeding their chicks.

Thrushes and many other birds may move further south and west to escape the cold weather creating huge influxes into the West Country, Ireland, France and Iberia. Indeed in Spanish the Lapwing is known as *Avefria* – 'Bird of the Frosts'. Britain is also likely to be invaded from the East by birds fleeing cold there and this creates increased competition for our local birds. There may also be regular movements, on a much smaller scale, by the tits in rural woodland. They may come down to the village gardens for peanuts and other food – often a matter of a few hundred metres but sometimes several kilometres.

The actions of the weather that threaten the birds are different for different species. The freezing of open water can trap birds in the ice and deprive diving birds of access to food. The birds may respond by keeping a small part of the surface free of ice simply by the pressure of numbers. They may move to areas where groundwater springs mean that the newly emerged water does not freeze. But many will move out and seek out new wintering quarters on the coast, where the salt water is unlikely to freeze, or in warmer areas. Frozen water can threaten the birds in other ways – for instance by allowing foxes and other predators to attack birds roosting in what would normally be safe places. The relative feeding behaviours of Grey Wagtails and Dippers on swift running streams determines the movement of the former away to unfrozen areas and the ability of the latter to stay. The Dipper can dive under the ice, which in these streams generally has air trapped under it, but the Grey Wagtail cannot and has to move.

Frozen soil and mud can both exclude the birds from their food and, in the case of many mobile invertebrates, force them deeper and out of range for the birds. In the estuaries freezing mud, or shallow water, can cause very serious deaths of waders immediately, or when their food is killed, later. The waders put on fat reserves for the winter which often become depleted through the winter and so late cold spells are often much more serious than early ones. Often only a minority of the birds dying are from British or Irish populations. Frozen soil can cause problems for seed-eating birds as the seeds may be stuck in or onto the substrate making prising them out an energy-consuming activity. Freezing rain or thawing frost re-freezing can be a real killer. This coats surfaces with a layer of ice and all sorts of birds may be affected. For instance the tits, Goldcrests and Treecreepers that would normally be searching fissures in the bark of trees are denied access to their food. In general small birds have much more difficulty keeping warm than large ones – they have a greater proportion of surface area and so cool quicker.

Snow itself can be a killer too. A deep snowfall in windless conditions can cover all available feeding areas for ground-feeding birds and so they will have to rely on the reserves of fat they have built up in better times. Snow, in blizzard conditions, is very unpleasant but the birds tend to shelter during the storm. They will be able to find open feeding areas when the immediate snowfall has stopped. After blizzards it is often the exposed hillside that the flocks of finches and buntings will resort to. Snow itself is an insulating blanket where it covers vegetation. Such species as Dartford Warbler, Dunnock and Wren may be very happy foraging below it: totally unseen. Surprisingly the snow itself is not a problem for owls who can hear the small mammals below it and take them unseen. For the Barn Owl rain is the worst hazard as they are unable to fly efficiently with wet wings. Hard weather does mean that the start of the breeding season, now for many species regularly initiated in late January, February and March, may be delayed and the breeding potential of the birds will be reduced.

Mild winters do allow more birds than normal to survive and can quickly make up for a really severe one. Small birds can recover their

Blackbird and Redwing.

numbers in three or four seasons and larger ones in half a dozen – even after really bad winters like 1962/63. However for most species the winter food supply is likely to be the limiting factor on population growth. One circumstance that goes against this is in winters when the forest trees produce a vast stock of acorns and mast. Then the tits and woodland finches do not have to find gardens for a plentiful supply of food and the rural woodland populations of these species may reach exceptional heights.

It would also be wrong to ignore the weather in other parts of the world where our birds are wintering. During the late 1960s and the mid 1980s drought conditions in the Sahel, the region just south of the Sahara, severely affected several of the species wintering there. Birds like the Sand Martin, Whitethroat and Sedge Warbler were very badly down after winters when the rains had all but failed. Particularly important is the area around Senegal, where a disproportionate number of our birds winter, and they were devastated. There have also been mortality incidents due to unseasonably cold weather affecting our Swallows in South Africa and, probably, our Swifts in East Africa. Seabirds may be influenced by what happens in West Africa (terns), South Africa (Storm Petrels), off Newfoundland (Kittiwakes) or Brazil (Manx Shearwaters).

Spring

Cold, wet or freezing weather in spring has a profound effect on the breeding season, as April and May are very important for most species. Nests which are lost or abandoned force the birds to go through the nesting process again. This may mean they lose out on a later brood through lack of time.

95

Many birds are prepared to give up quite easily if they are not far advanced with the breeding attempt. However the time invested, and the time which will thus be 'lost', mounts up as the breeding attempt progresses and the adults are more reluctant to give up. Nidicolous birds (those whose young stay in the nest) are particularly vulnerable when the young have hatched. The chicks have to be fed and often need special foods to grow properly – invertebrates from normally seed-eating parents for instance. However in cold weather they also need to be kept warm, by the brooding female, for several days after they have hatched. The parents are in a classic 'Catch 22' dilemma. Brood the young and they may starve – feed them and they may die of hypothermia. It is even worse if it is wet: the damp parent may cause the brood to cool rather than warm! Nidifugous birds (the young leave the nest immediately) have just as bad a time, particularly in wet weather when many chicks die of cold.

The birds are often feeding on seasonal insects, either as adults or, more often, as larvae. They must second guess when the food supply is going to be at its best. This means that they have to lay the clutch and start incubating at the right time for the peak of caterpillars – in the case of the woodland tits – to be roughly three weeks later. If things go wrong the productivity suffers but if they get things right huge broods may be raised – no losses even from 15 and 16 egg clutches. Huge numbers of caterpillars hatch, but a high percentage may die if there are late frosts. Heavy rain, often a feature of our summers, can wash the insects off the trees onto the ground, causing additional problems. Rising water levels may wash out the nests of birds like wader and gulls – and high tides can be a problem too.

Once more we have also to pay attention to the weather elsewhere in the world. In the spring the most crucial area for our returning summer migrants is the Mediterranean. They pass through the area having had to cross the Sahara. If the weather is bad as they arrive they may die, but will certainly have to replenish depleted resources in difficult circumstances. This means that they are delayed and hence the breeding season in Britain may be days or weeks late. Sometimes weather affects the birds on migration in the Mediterranean. For example, the male Willow Warblers on their way to Southern England at the start of the passage might be caught by a cold spell and suffer severe mortality. But their females, and both sexes heading for the North of England and Scotland, passing through later, would be totally unaffected if the weather improved, and neither would suffer losses or delay.

Spring effects are very likely to be passed on to the breeding population levels in later years as poor numbers of young fledged in one year will mean that there are fewer birds that can join, or recruit to, the breeding population the next year. This effect is immediate for small birds that breed in their first year but is masked, and probably undetectable, for larger species which delay first breeding for several years. For instance it is quite normal for Tawny Owls in some areas to fail to breed every four years or so when the voles, which form their food supply, suffer population crashes. However the adult birds have a high survival rate and the young birds are likely not to breed immediately so the hiccups in population level are smoothed out. However after a year when they have missed the caterpillar peak – or the caterpillars have failed – there is an immediate drop in breeding tit numbers that is made up as soon as there are a couple of normal years.

After good breeding seasons the numbers of young birds may be very high indeed. This will put additional pressure on the resources and recruitment into the breeding population is seldom as high as might be expected. The additional juvenile and first winter birds do serve as a cushion against winter losses but these birds almost always survive much less well than the adults. For small birds the average pair of adults with half a dozen young will normally mean, if the population is stable, that six out of the eight birds will die so that just two breeding birds are alive for the next breeding season. One of these is likely to be one of the original pair and only one will be from the six youngsters. In bad seasons the effects are even more marked. For Sand Martins the population crashed because of the Sahel drought over the winter of 1968/69. For that year I calculated that adult mortality was as high as 82% compared with 65% in a normal year, but for the young birds it was 96% compared with 77%! That is the adult survival rate was almost halved but only one in eight of the young birds that might normally have been expected back returned. The decrease in breeding population over the one winter was about 70%.

Summer

Bad summer weather can be equally difficult for the breeding birds and severely damage the species' prospects in later years. Different extremes can affect different birds in different ways. Very hot weather may be very good for insectivorous species but it bakes the ground hard. This means that thrushes will stop breeding much earlier than in damp years. It also means that the survival rate of young Rooks is much reduced, as they are unable to extract soil invertebrates. Wet weather will mean that the thrushes are able to extend their breeding season but the smaller insectivorous birds may have a hard time. Ground-nesting species are vulnerable to flooding and water-logging and the aerial insect feeders – Swifts, Swallows and martins – are in for a hard time. However the Swallows and House Martins should not have too much difficulty finding mud for their nests!

Windy weather is not, generally, a problem for most species except aerial feeders. However wind on the surface of water creates ripples and the birds that dive, direct from flight, for their prey may get into serious trouble. Mass failures of breeding seabirds have been attributed to this although other factors are often implicated too (see Marine and Coastal). In exceptional circumstances nests may be washed off ledges by very high seas and this is the reason why seabirds are unable to establish regular breeding on Rockall. For land birds summer gales do destroy some nests. One species to suffer particularly badly was the rare Golden Oriole. These birds nest in the outer twigs of poplar trees in the Fenland area and, recently, they lost the majority of their nests after a mid-June gale.

Summer weather seldom puts the adult birds at risk but it may reduce the nest productivity, the number of nesting attempts and the survival of the young birds that have already fledged. For many birds there are, routinely, huge losses of young birds between fledging and recruitment to the breeding population so deaths of newly fledged young perhaps 'do not matter'. But our understanding of just which young birds are important is poor and this belief that they do not 'matter' is based on the assumption that other young birds will survive. The parent birds of many species look after their young for some time after they leave the nest but, in multi-brooded species, this cannot be for more than a few days.

Autumn

This is the time of 'mists and mellow fruitfulness' when most birds have plenty of food, long hours to find it and mild and untaxing weather. This is just as well as the adult birds have to renew their plumage and, if they are migrants, prepare for their journey. The young birds also have to go through a moult, though usually not so extensive as their parents, learn how to survive and prepare to disperse or migrate. Wet weather can actually be a boon to many species as, if it disrupts the harvest, spoilt crops might be available for the birds. Similarly winds might cause top-fruit to fall to the ground and lie there as windfalls for the thrushes to eat weeks or months later. However autumn winds can also shake the seeds off trees like birches so that they are not available for finches like Redpolls and Siskins later in the year. Bigger seeds, like beech-mast and acorns, blown off the trees often stay on the ground for months and are fed on by many birds and small (and big) mammals.

Autumn is, of course, the season for migration and dispersal. The short dispersal movements of resident species are not often hazardous but even roosting flights have their hidden dangers. In foggy weather Starlings may be attracted to lights and collide with buildings, windows, wires or even the ground! Migrants are also prone to losing their way and colliding with objects but the new electronically flashing lights on lighthouses are much safer than the slowly rotating, and now obsolete, occluding system. Young birds are prone to getting lost and may die as a result. This is sometimes due to weather but more often to inexperience.

The future

Global warming does not mean that the temperature in particular places will become warmer, just that the average over the world will rise. However current predictions are that Britain and Ireland will warm. Most scientists working in this field believe that the weather will also have more extreme episodes – like the famous hurricane that struck Southern England. Certainly there have been some exceptional episodes of flooding as well as gales and there are predictions, from some computer models, that there will be wetter (or snowier) winters and drier summers. Experience over the last ten or twenty years certainly shows that many of the warmest years since records began have happened recently. However within this record has been a worrying number of cold (and wet) late springs in Britain and Ireland and inclement weather during the spring migration for our birds through the Mediterranean. The likely effects on our birds are many and varied. For instance the majority of the birds breeding in Britain are now laying earlier than they were in 1971. The first determination that this was happening came from the analysis of 65,000 nest record cards of 65 species from 1971 to 1995. On average the difference is about ten days and the 20 species for which the results were significant are:

Oystercatcher	8 days
Curlew	8 days
Redshank	11 days
Tree Pipit	8 days
Dipper	8 days
Wren	9 days since 1980
Redstart	11 days
Whitethroat	9 days
Blackcap	7 days
Wood Warbler	5 days
Chiffchaff	14 days
Willow Warbler	6 days
Long-tailed Tit	11 days since 1979
Nuthatch	6 days
Magpie	17 days
Carrion Crow	7 days
Starling	4 days since 1983
Chaffinch	6 days
Greenfinch	8 days
Corn Bunting	16 days

Carrion Crow.

Only three were laying later and this result was only significant for the Stock Dove, a species with a very long nesting period. This trend can certainly continue if the springs get earlier and earlier. However the actual effect on bird populations may or may not be important. If the birds have the same number of breeding attempts and are as successful as they were before there will be no effect. If the birds survive the winter better, and are able to breed more successfully, it is possible the background population level may increase.

The earlier springs have their effect on all sorts of other biological events. Earlier spawning of frogs, earlier flowering and bud burst of plants and earlier emergence of insects are already being recorded. But the problem is that the birds may use different clues to trigger breeding than the insects, for example, are tuned to. It is possible that the tits' laying will get out of step with caterpillar development. Certainly, when this happened over much of England, Scotland and Wales, 1998 was the worst breeding season on record for Blue and Great Tit. Such disruption of the fine tuning brought about by many generations of evolution may take many generations to put right. During this time it is likely that the carrying population of the bird species will be reduced.

The fine, hot summers may alter the growing seasons for our plants. If they have spring and autumn growing seasons, with an interval of baking hot weather with no new plant growth, it would hasten the end of the breeding season. The same summer weather would put at risk many species that require wetland habitats which would quickly shrink. Areas which become suitable for breeding could well become much too dry before the birds were able to complete a breeding attempt. The farmland habitat is likely to change quickly and there may be new crops. The country could resemble the Mediterranean, with farmers planting lots of maize and sunflower. Dry fields with olives and almond trees to provide shade for the crops and acres and acres of vines! Clearly the avifauna would be very different. If the warming happens quickly there is likely to be a hiatus between the native trees in Britain suffering and the species able to resist the heat growing. Changes on this scale are clearly very worrying.

The drying out of freshwater areas – lakes, ponds, streams, marshes, reed-beds and fens – will be a real problem for wetland birds. We have already lost vast areas of lowland marsh and water meadow as a result of drainage. Many areas have lost their Redshank and Common Snipe populations already and these species are particularly at risk. However there are now large numbers of conservation organisations who own and manage these wetland areas for the benefit of wildlife. So long as the winter water is available to regulate the summer water levels these sites are likely to be safe. Working water bodies – reservoirs – are likely to be subject to greater fluctuations in water level than they were before. This is not good news for waterside birds who may start nesting in a wet area that rapidly dries out. Even damp pasture is in danger of drying out and becoming unsuitable for breeding waders like Lapwing and Curlew.

One likely effect of warming is that the arctic-alpine species of birds found in the Highlands of Scotland are likely to be lost – birds like Ptarmigan, Dotterel, Snow Bunting, Red-throated Diver and several other waders could go. However there are models which seem to indicate that the warming is likely not to have a very severe effect on the habitats these birds need and it could be that they may be reduced in numbers but still hang on. We are certainly receiving new birds from the South – Cetti's Warbler and Little Egret, now breeding in Southern England and the Irish Republic – even Hoopoe and Serin. Other birds are moving north and west – Reed Warblers are in Scotland and in several Irish reed beds, Nuthatches are breeding 30 or 40 kilometres further north than

Hoopoe.

Dartford Warbler.

they were 25 years ago and Little Ringed Plover may now number as many as 1,000 pairs in 50 years since they arrived. Wood Warblers and Pied Flycatchers are now breeding regularly in Ireland having colonised woodland that has been in existence for a very long time.

A feature of the future may well be that the really long, severe winters that have happened about every ten or twenty years may become a thing of the past. This will probably smooth out the population figures for the vulnerable species. Birds like the Grey Heron and Kingfisher whose food is directly denied them in prolonged frosts will no longer be subject to declines of two-thirds or more. Many small birds are now irregularly reduced by 50%, 70% or even 90%. If this does not happen the birds are unlikely to increase in numbers – simply they will not crash in population. The birds that are doing well are those that are very vulnerable: the Dartford Warbler was reduced to 11 pairs after the 1962/63 cold winter. The birds have been able to re-occupy areas where they became extinct and are colonising new areas – such as the Suffolk coastal heathlands – and there have been over 1,000 pairs for several years. Another warbler, Cetti's, that newly colonised England less than 30 years ago also stays for the winter. These birds have increased, fairly steadily, to over 500 singing males and have occupied their special habitat over much of the southern half of England and Wales.

One of the most worrying features of climate change is the predicted rise in the global sea level. This is caused by the melting of the Arctic and Antarctic ice and of glaciers and also of the expansion of sea-water as its temperature rises. The predicted rise – of 30-50 cm in a few decades – does not seem to be much. After all the tidal rise and fall in some parts of Britain is as much as 17 metres! However even such a seemingly small change will have huge consequences for the coastal ecology with the predicted loss of much of our coastal mud-flats and salt marsh. Normally this would be replaced by the sea inundating new land but this is now defended by hard coastal defences. 'Managed retreat' is the buzzword and this would mean that areas that were previously defended will be allowed to flood. Thus our coastal habitats would be replaced. Unfortunately there are many areas where the land inland of the estuary is developed and even where it is farmland; there will be huge social pressures to keep up traditional defence. Our salt-marsh and mudflats are not major breeding areas for our birds but many of them, and a lot more from abroad, are used by birds during the winter.

Perhaps these different scenarios, and the influence on our bird populations may seem far fetched and rather drastic. If the extent of global warming being predicted at the moment by some teams of climatologists is correct, these, and other effects, will surely come to pass. Changes are even happening because the increased levels of carbon dioxide are altering the subtle balance between different species of plant. For instance some of the increase in productivity of wheat in the last few decades has been because of anthropogenic carbon dioxide – it is not just the climate and weather that have changed because of the greenhouse gases.

Burton, J.F. 1995 *Birds and Climate Change*. Christopher Helm: A.C. Black.
Crick, H.Q.P., Dudley, C., Glue, D.E. & Thomson, D.L. 1997 ' UK birds are laying eggs earlier' *Nature*: **388**, 526.
Crick, H.Q.P. & Sparks, T.H. 1999 'Climate change related to egg-laying trends' *Nature*: **399**; 423-424.
Elkins. N. 1983 *Weather and bird behaviour*. T & AD Poyser.
Moss, S. 1998 ' Predicting the effects of global climate change on Britain's birds' *British Birds*: **91**, 307-324.
Thomas, C.D. & Lennon, J.J. 1999 'Birds extend their range northwards' *Nature*: **399**, 213.
Toomer, D. (ed) 1999 'A bright future?' Issue 223 of *BTO News*.

PERSECUTION AND PROTECTION

This is the yin and yang of man's relations with birds. On the one hand the sheep farmer loathes the Raven which he thinks kills his lambs, on the other the birder from the distant metropolis is thrilled by its mastery of the air. In fact we are all really exploiting the birds, even as watchers, but those interested in their conservation will not wish to harm them. They will hope that they are properly protected and their habitat is preserved. Others may wish to conserve the birds but exploit them very directly – by shooting them! Their attitude to the birds they wish to shoot should be, and very often is, very like that of the outright conservationist. The last thing they want to do is to hazard the future health, let alone existence, of the populations of their quarry species.

Others have different agendas. A minority of gamekeepers are keen to destroy birds of prey (illegally) and corvids (legally). Some people feel, wrongly, that Sparrowhawks are the driving force leading to the small bird declines and should be culled. Racing pigeons fanciers are often are against Peregrines. Many anglers feel that Cormorants and Goosanders are unfair competition for 'their' fish, farmers see the crop being eaten by Woodpigeons as 'too much' and there are many gardeners who despair of keeping the birds off the soft fruit or cherries they grow for themselves.

These interactions are the subject of this chapter. They are, of course, regulated by law and the legal framework for bird protection and nature conservation is changing quite rapidly. Indeed, apart from the game laws, there were hardly any regulations 100 ago and nature reserves did not get going until about 60 years ago. The extent and speed with which the 'industry' of birdwatching has accelerated can be gauged by the membership of the RSPB – a few thousand in 1960, over a million by 1998. The society's magazine had one ad in it 40 years ago for a bird holiday (bed and breakfast accommodation at an East Coast cottage) now *Birds* advertises pages and pages of holidays all over the world. Indeed there are more than 500 holiday companies and holiday properties advertised in the most recent issue. Just one company advertises over

150 different overseas tours in one year with prices up to £4,875!

This vast industry, with huge numbers of books, tapes, CDs, videos, TV programmes and so on, illustrates the degree of interest within the country for birds. It is this that leads the legal and cultural climate which is 'pro bird'. The people that are against birds are almost always only against the particular birds that they feel are acting against their interests. It might be the Starlings nesting in the roof that wake them up too early in the morning, the Woodpigeons that eat their vegetables or the Carrion Crow that spends the breeding season rapping against the windows. In these situations there are steps that can be taken to alleviate the perceived nuisance – short of harming the birds – and in most cases this is what will happen.

Birds as food – and other direct exploitation

Many people consider that various wild birds are delicious! Their direct exploitation of the birds has been regulated for many years by law and some species are important, economically, within the countryside. The Pheasant in farmland is the paramount example. Birds are reared and released for shooting in an industry worth many millions of pounds and this is directly, but not deliberately, crucial to the populations of other birds on farmland. There is a greater biomass of Pheasants in the autumn in Britain and Ireland, in the wild, than any other species. They weigh in at about 25,000 tonnes and the next species is Woodpigeon at between 5,000 and 10,000 tonnes. There is no suggestion that the populations of either species are being put at risk by exploitation by shooting. Both species are among the short list of farmland birds that are increasing.

Red-legged Partridge.

This was not always the case in the past. In historic times such rare species as Crane, Spoonbill, Dotterel, Ruff (and Reeve) and Great Auk have been hunted and eaten. At the time there were no laws to control the slaughter and the concept of conservation had probably not crossed the minds of the hunters. The Great Auk was rendered extinct and, apart from the Dotterel, the other species ceased to breed in Britain. The Dotterel is a very interesting case as there are several 'Dotterel Farms' scattered over the chalky high ground in Cambridgeshire near Royston. There are many records in Victorian times of the birds being shot and trapped in the area and then being sent to Leadenhall Market – the traditional poultry market in London. This was more recently thought to be strange and a thing of the past. However in the mid-1950s Frank Pepper, a farmer interested in birds, found them in the same area. It has now been established that there is a regular passage in the first half of May. Needless to say the birds are now watched and not eaten!

Earlier, at the end of the 19th century, the chalk downland of Sussex in particular also figured largely in a trade in small birds – mostly dead for the table. These were mainly Skylarks but also Wheatears, presumably on passage, and many other species. They were trapped by shepherds, to earn themselves extra money, and sent in bundles of a dozen birds to Leadenhall Market. Apparently hundreds of dozens of birds would be sold each day! Many were exported to France where the Skylark is still hunted to this day. Presumably the trade relied on the swiftness of the railway to ensure that the birds were still edible when they arrived at their destination. There would be no chance of catching this number of birds at present but it is not likely that the trade itself is responsible for the decline. Later the bird collectors became very interested in rare birds caught and paid over the odds for good specimens for their cabinets. Dealers became involved and there was much deception as birds caught abroad were passed off as having come from the catchers. The history of the *Hasting's Rarities* makes fascinating reading but does not impinge on our breeding populations.

Current bird watchers may be surprised to know that common birds like Rooks and sparrows were eaten (Rook pie made from young birds just before fledging). Plover's eggs (supposedly Lapwing's but often Black-headed Gull's) were a delicacy until recently – indeed the gull alternative is possibly still legal. Moorhens were supposedly a delicacy for travellers – baked in clay over an open fire, they tasted like chicken. There was even the tragedy, about 25 years ago, at one of the sites in Herefordshire where I used to ring Pied Flycatchers. One year the brood of Ravens, very near fledging, ended their lives in an SAS cooking pot. The squad was learning survival techniques and the owner of the site, very proud of his Ravens, was not best pleased! Many will realise that there are bill marked Mute Swans on the Thames. The simple nicks show that they belong to one of two City Guilds (Vintners or Dyers) and that all the rest belong to the Crown. What they may not realise is that this is because Mute Swans used to be a valuable commodity for use as the centrepiece in any high class banquet. Indeed the species was so tamed during mediaeval times that it was only the remote birds that could be classed as wild. A very complicated system, involving bill marking, regulated the ownership of the lowland birds.

Many populations of seabirds were exploited for food from time immemorial. The whole population of St Kilda, off the West Coast of Scotland, and many others were dependant on 'sea-fowle'. They were harvested in huge numbers but with some thought of sustaining the populations for later years. Young birds were particularly valuable as they were often very fat before fledging. Fulmars, Puffins and Gannets were at risk and they were not just eaten. Some were rendered down for their oil and the feathers from all of them were collected for use or sale. The men from the parish of Ness, on Lewis, are still allowed to take young Gannets for food from Sule Sgeir. These youngsters are called *gugas* and their current licence is for 2,000 each year – the colony is apparently stable. Other seabirds, apart from gulls, were regularly egged. The 'climmers' at Flamborough Head (Yorkshire) used to be let down on ropes to take Guillemot eggs for food. In Victorian times they were paid a retainer by egg collectors for first choice of the interesting variants brought up. These activities killed many birds but it is unlikely that the effect can still be seen on any populations – save the lack of increase in the numbers of Gannets on Sule Sgeir. However the same cannot be said of the destruction of the Guillemot colonies along the chalk cliffs of Beachy Head on the Sussex Coast These were shot out during Victorian times by 'sportsmen' who had special summer excursions, by train, laid on from London! The last straw was some massive cliff falls and, to this day, these cliffs are aukless.

Little and Great Crested Grebe.

Now there are rather few quarry species that are regularly eaten – apart from the game-birds and wildfowl. Woodpigeons are partly shot to be eaten, but primarily to protect crops from their attention. Snipe, Woodcock and Golden Plover are shot and eaten but their numbers are in decline and many conservationists, and not a few sportsmen, are concerned by the current trends. The detailed situation with regards to shooting of particular species and groups is to be found in the chapters dealing with their principal habitats. Ffor instance the grouse are dealt under Uplands and Woodland. The side effects of gamekeeping activities in the persecution of birds of prey – now illegal – was pretty awful but is much better now. There is no suggestion that culling of corvids is leading to a decline in their numbers. Indeed Magpie, Jackdaw and Carrion Crow are again amongst the species on farmland which are increasing apace.

Of course birds are also exploited, to their detriment, for all sorts of other reasons. The humble Wren used to be hunted at Christmas in a weird bit of folksy behaviour which spread through much of the Celtic world. This was possibly a distant memory of the bird nesting in the caves where the people lived. Feathers were collected from various species – Great Crested Grebes suffered greatly – and some were shot for fashion requisites. This was the reason that the RSPB was founded. Specimens of Kingfishers, Barn Owls, raptors, all sorts of birds were stuffed and sold as decorative items for the cottage mantelpiece and really rare birds (and their eggs) were collected and traded. There is no doubt the Osprey and White-tailed Eagle were exterminated by collectors and they have not helped Red-Backed Shrikes – even though they have been protected for many years. There is even a trade in feathers for tying the flies used in fly fishing.

In some ways the saddest exploitation of birds is for the caged bird trade. Almost all the people who used to (and do) keep birds in cages would class themselves as *birdlovers*. Conservationists would not agree with them and consider that birds should be wild and free. The current scale of British bird catching for caging, illegal for many years, is difficult to quantify. A recent case concerned illegal exports of birds, mainly finches, to Malta. It was said to have involved tens of thousands of birds and was worth many hundreds of thousands of pounds. The scale of the trade in Victorian times was huge. Mayhew's *London Labour and the London Poor* records the trade in the mid-1800s in the metropolitan area alone. Most of the birds came from local catchers in areas now under concrete. The annual figures taken are amazing:

Linnet	70,000
Goldfinch	70,000
Skylark (and other 'larks')	60,000
Song Thrush	35,000
Bullfinch	30,000
Chaffinch	15,000
Greenfinch	7,500
Blackbird	7,000
Robin	3,000
Nightingale	1,000

In addition 215,000 'edible larks' were taken from further afield. At that time the downs of Cambridgeshire and Bedfordshire, possibly because the railway transport from Sussex and the South Downs had not yet developed, were the main centre for catches. There were even boys selling nests with clutches of eggs for raising under cage birds! The birds could be very valuable if 'trained' – that is singing in cages without being scared. Half a crown might be got for a good finch and 15/- (75p – but then much more than a week's wage) for a singing Nightingale. Freshly caught finches might fetch as little as two or three pence (1p). Most of the birds were netted or trapped but some were reared from the nest. The consumption of small birds by London was thus half a million or more. And of those destined for cages about a half died before they were sold. Robin is low on the list as there were all sorts of superstitions against keeping them in cages. However a good songster could fetch £1 at a time when a decent weekly wage was half that!

Birds behaving badly

There are all sorts of ways that birds might be said to be *behaving badly*. That is doing something which some people might think was wrong –

against their interests. However those activities that are recognised by society as being actionable are very carefully restricted. The rule is that all birds are protected but that there is a general licence that allows, in certain circumstances, the 'control' of certain common birds. These used to be classed as 'vermin' and include many corvids, some of the gulls, Woodpigeon, Starling, etc. They may be controlled by defined legal means, if they are:

1) Hazarding health and safety – this might be transmitting a disease or crossing the flight path of an airfield.

2) Acting against the conservation of other wild birds – presumably more 'valuable' wild bird populations. For instance a Carrion Crow raiding a Little Tern colony.

3) Seriously damaging crops, livestock, fisheries etc. – this allows the control of crows raiding the nests of gamebirds or Cormorants at fisheries if real damage can be proved.

Health and safety can be quite clear – faeces under Starling roosts is unpleasant, may spread disease and cause people to slip on slimy pavements. However the legal restrictions on control mean that few roosts are subject to control and, in most cases, the action would be to try to disperse the roost and clean up the mess. Sadly Starlings are in decline and this is a problem also in decline. There are real problems round airfields where birds can bring down aircraft involved in birdstrikes. However few birds are actually killed to prevent accidents: ecological measures – like allowing the grass to grow long – and scaring techniques are now used.

Birds very often eat other birds and the conservation clause is sometimes invoked. Carrion and Hooded Crows can be very damaging at colonies of ground nesting birds and legal killing is allowed in these circumstances. In the case of species other than the common ones, where control is allowed under the general licence, other measures are often taken. Two Little Tern examples illustrate this quite neatly. In Norfolk, near Yarmouth, a Kestrel found the terns a useful food supply and the wardens lost large numbers of birds. In subsequent years the Kestrels were given an alternative food supply – dead mice – close to their nests and did not look the gift mice in the mouth! Later the tern chicks were also supplied with neat little shelters under which they could hide if their parents made the alarm calls signifying the approach of a Kestrel. The success of these measures was very gratifying and hundreds of young terns have bFeen raised each year where it used to be just a few dozen!

One contentious bird is the Magpie. It is legal, under the general licence, to kill these birds to protect the small birds they eat in the breeding season. Some fanatical Magpie-slayers spend their whole time

Sparrowhawk chasing Linnets.

doing this. One who contacted me had killed about a thousand a year for several years. However the wide area he was trapping in put this into perspective – he was responsible for a few percent of the Magpie deaths that happened each year in his area. It is quite likely that his trapping has no effect on the overall population and the only 'good' he is doing is to his self esteem.

Proof of damage to crops and livestock may seem to be easy. Crop growing in field, birds come to feed on crop, damage proved! However this is not always the case and the birds may be eating just that part of the crop which cannot be harvested or, earlier in the year, they may even be encouraging the plant to produce more shoots and an enhanced harvest. Obvious damage can be done by many species – for instance Woodpigeons on ripening grain, winter brassicas or clover. In winter there are many areas where migrant geese congregate to feed on crops or waste. Pink-footed Geese guzzle carrots in Lancashire and many species nibble and tug cereal and grass plants but birds eating left-over potatoes or sugar-beet tops are not a problem. However there are great difficulties over the potential damage done to fish stock by Cormorants and other fish-eating birds. They certainly eat some fish but they may actually simply make the fish more difficult to catch for the angler! Indeed in many situations minimal, if any, attempts are made to provide the fish with any ecological advantages over the potential predators. Natural waters will have plenty of weed and often rocky areas or extensive tree roots where the fish can hide. Put and take fisheries are often featureless goldfish bowls on a gigantic scale *inviting* the predators to come and feed. Even the most recent research does not indicate the *actual* damage to fisheries by Cormorants can be proved to be widespread. In these circumstances it would be a travesty if the birds were killed when the basics of the fishery were wrong. Most licences would only be granted, if ever, to allow minimal destruction of the birds integrated into a well researched scaring programme.

There are no provisions, within the law, for other reasons for killing birds. For example the pair of Carrion Crows that wreaked a path of destruction costing well over £100,000 amongst new cars parked in the North-east in a large field were immune. Their tearing off wiper blades, various items of trim and actually destroying windscreen seals was allowed to continue unabated as the distributors would have been killing them to protect property and not crops! Birds that do not fall under the general licence can be killed, under a specially issued licence, but these are seldom issued. 'Messy' House Martins nesting on a building should be safe – but the nest can be (and sadly often is) knocked off in the winter. However, by and large, birds who do behave badly may be killed under certain circumstances but the level of killing is not now likely to have any effect on the national population.

Bird protection

Legal protection of birds in Britain and Ireland is good – and is pretty well enforced. The birds can generally go about their business without casual acts of vandalism or cruelty affecting them, their nests or their young. There are cases every year where prosecutions succeed where the law is broken through ignorance or wilfulness. It is even an offence to disturb the rarest species whilst they are nesting. There are few people who think they are allowed to shoot a Barn Owl, to stuff, or a Tawny Owl because it keeps them awake. However it was not always so. Disturbingly recently one could find species that are now protected on many a gamekeeper's gibbet in Britain and, much less than 20 years ago, one could see the dread sign *Poisoned Land* in the Republic of Ireland. This meant that poison baits were being put out to kill foxes or crows – or anything else. The signs were supposed to warn people that their dogs were at risk if they entered! Perhaps it was the efforts of the Irish Taoiseach (Prime Minister) to introduce White-tailed Eagles that caused this barbarous practice to be stopped. This has allowed the Buzzards that breed in Northern Ireland to spread into the South and consolidate their population.

The idea that birds are protected and should be allowed to live alongside man is new. Such an attitude towards other creatures was not apparent in most Christian societies until recently. Perhaps St Francis was one of the first people to start things off but there are others. For instance St Serf, on his island in Loch Leven, had a tame Robin some 1,250 years ago. The interest in birds can also be traced through the history of garden bird feeding and seen in the attitude of the birds to people. British and Irish Robins are much tamer than ones in France.

Protection probably started as a rather lackadaisical 'live and let live' attitude, developed into a more caring mode and is now the basic assumption amongst most of the public. The attitude of people interested in birds has also developed similarly. Victorian, and earlier, ornithologists would have looked upon their armoury of guns in exactly the same way that their successors, 100 years later, look upon their binoculars and telescopes. 'What's missed is mystery, what's hit is history' was the motto and this attitude was still alive as recently as 35 years ago amongst a few. However, ultimately based on the specimens in collections, we now have excellent books of identification and, using modern optical design, equipment that brings the birds into focus at a great distance. This extra dimension has added to the interest, and concern, for our birds. Add the popularity of TV programmes about birds, conservation, extinction, evolution, etc., and there is little reason to enquire further.

However the protection afforded our birds does sometimes break down when they migrate away from our shores. Even within Europe there are birds that we protect assiduously that are legally hunted in other countries – for instance the Skylark. Rather few of our Skylarks venture to the areas where they are likely to be killed but such was not the case with some area species. The Montagu's Harrier that were just holding on in Britain and Ireland in the middle of the century (about 30 pairs) suffered cruelly from the shooters in France. When the populations started to build up again recently I was waiting for the first recovery in France by a hunter to really lay into them – the species was by then properly protected. The recovery came in but it illustrated totally natural predation – the corpse was found in the nest of an Eagle Owl! Protection for the Turtle Dove, traditional quarry in Portugal, has now been established there. The warblers and small birds passing through this area are probably still trapped, in a small way, but are seldom reported as the trappers know it is against the law. This trapping was traditional and probably did no lasting damage.

Outside Europe the situation is rather murky. Subsistence trapping for food is not likely to lead to many ringing recoveries and may well, otherwise, be unreported. A few years ago a massive Swallow roost in Nigeria was found to be being exploited by the local villagers for food. British ringed birds were involved and an agreement, involving a generator and advice about an alternative source of village reared protein stopped the deaths. Such practical solutions and the growth of legal protection reinforce each other. One worrying use of our birds that may still be going on is for bait for long-line fishing. This used to happen with Kittiwakes on the Grand Banks off Newfoundland but this hazard has ceased. The fishery is closed because the cod stocks are so low. Similar exploitation may continue in the Bay of Biscay but regulation of fishing is so much more of a political issue now that it may have effectively been stopped.

Bird and nature reserves

It is amazing to think that the extensive and varied suite of bird and nature reserves, scattered over Britain and Ireland, from Connemara in the West to Minsmere in the East and from Scilly in the South to Noss in the North, has effectively been constructed in the last 50 years. The very thought that even a small area of land should be controlled and managed for the good of the birds and other wildlife would have been considered most peculiar 100 years ago. There were individual, eccentric, landowners who wanted to manage the estates they owned for the benefit of the natural beauty in areas like the Lake District or the Highlands. Very many more were interested in the sporting use of the land but, often, this was only for the body count of the birds that could be shot.

Bird and nature reserves are most often associated with particularly rare and interesting habitats, and rare and interesting species. These habitats, and birds, tended to be in poor areas of land unvalued for agriculture and undeveloped so that large areas could be bought cheaply. Whether they were areas of remote uplands or wetlands close to centres of population the explosion of new reserves, big and small, was staggering. Now, with much more money available, new and huge reserves are being bought every year and smaller ones come into being almost every week. The extra money now reflects not just the increased interest but also increased competition for the sites. For instance remote areas of peatlands might be artificially valuable as potential forestry because of the tax regime. The same sort of areas, in Somerset or Yorkshire, are valuable for their peat which, if milled and bagged, will be a multi-million pound commodity for gardeners.

What is most exciting is the idea that man's ingratitude to nature should be reversed and there are new reserves planned which flood drained land – as at Lakenheath in Suffolk – so returning farmland to land dedicated to wildlife. For many people areas like golf courses seem like excellent wildlife areas; some are, but many others are artificial and also have a high use of pesticides to ensure the good surface for the golfers. Conservationists in the Republic of Ireland are still fighting current proposals for the development of golf courses on internationally important dune systems in the West of the country – for example in Co. Clare. However golf courses carved out of farmland can be a great gain for wildlife. As we have seen the integration of wildlife interest with growing timber can greatly enhance bird populations. Indeed there are several species of woodland birds that are increasing in population at the moment – and for the future. All is not as easy now since there are other uses for woodland – paint ball games, off road driving, orienteering – that may be bad for the birds. Recently illegal biking has reduced an SSSI (Site of Special Scientific Interest) in Berkshire to little more than a series of muddy and eroding tracks. It should be lowland heathland with such species as Dartford Warbler, Woodlark and Nightjar!

This actually points up one of the real issues. Nature reserves are marvellous but they are not necessarily reserves. Even statutory reserves – like SSSIs – are likely to be damaged by mistake, deliberately by the owners or by government, even against the advice of their own advisers. For instance Cardiff Bay was an SSSI and held a good population of intertidal waders. It was not what some ignorant people thought would have been the right back-cloth for a modern development and so it has been dammed and turned into yet another freshwater area. Scandalous! One hopes that the Irish will be more sensible and throw out the proposals for a similar scheme on some Dublin Bay mudflats. The same thing has happened in other estuaries for industrial use such as Seal Sands on the Tees and around Felixstowe in South-west Suffolk. These sites are actually not safe against national need or commercial exploitation. They are often under attack and must be defended, time after time, through the planning system – at great cost to the conservation movement. Something has to be done about this so that, at the very least, there is a moratorium on successive planning applications, for very similar developments.

However one must note that few rare birds that breed in small or medium sized territories in the sort of habitat that make up reserves, that are not doing well. Indeed there are probably several with all the normal breeding population living on reserves where they are looked after well. They are even being encouraged to spread into new areas – again managed with their needs in mind – with some success. Birds nesting in reed-beds, gravel pits and other wet habitats are particularly well suited. There are many pressures on holes in the ground. When planning legislation started it was expected that newly dug gravel workings were to return to agricultural use after they had been exhausted. This was expensive and not cost effective – the land was often marginal and thus not valuable for farming. Then the holes were filled with water and became fishing lakes, lidos, sailing lakes, water skiing venues and even nature reserves. These are cheaper ways of dealing with the redundant pits and often provided income for the owner. However for many years holes in the ground have become very valuable commodities as places where waste can be buried – and nature reserves are looked upon as a very pleasant alternative use to that by the local people.

Even rare birds with wider distribution are doing well as more and more people get interested in them. Who would have thought 70 years ago that the Red Kite would not only have pulled through but would be increasing apace. In the mid-1930s the Welsh population went down to one breeding female and the birds were in a parlous state until about 25 years ago when new blood, a female probably from the German population, joined them and suddenly brood sizes of two and three appeared. By 1992 there were more than 100 breeding pairs and there were 1.6 young reared per successful pair. In 1989 the decision was taken to give them a helping hand and Red Kites from Spain and Sweden were introduced successfully to England and Scotland. The introduced birds are doing very well too and will soon overtake the 'native' population – their brood size has consistently been well over two chicks per successful breeding pair.

The future

There is no chance that the present overall protection from persecution will be reversed. The attempt by some factions to turn the clock back and revert to Victorian values with birds like Sparrowhawk, Hen Harrier and Peregrine being shot, to preserve other birds, particularly game species and racing pigeons, is currently unthinkable. One reason is that the basis of our own laws is European legislation. Even the licensed removal of particular individuals is not currently an option. If we were to try to get the European law changed Britain would be unable to resist other changes to the bird protection laws which would be very detrimental. There is possibly a concerted attempt to push these changes through but the research being done on the problem has removed the global reason for the change – the general declines in small birds are not to do with predator birds. The other reasons – the influence of predation on the numbers of gamebirds and racing pigeons – are looked upon as very factional and, with so many people interested in wild birds, are very unlikely to gain approval. It is up to us to make sure that they are thwarted.

Finally the protection of habitats and reserves is by no means perfect but the amount of money and effort being channelled to reserves is high. I have little worry about the health of bird populations that are based on reserves as their 'home ground' is in good hands. However one has to say that there are grave worries about those species that are in the wider countryside – or use the wider countryside outside the breeding season. There is little that we can do, at the moment, to stop the continued deterioration of these general habitats without, for instance, real help from the CAP. The announcement of modulation in December 1999 means that this will come sooner than I had expected. The real question is whether this will be soon enough and strong enough to save some of the supposedly commoner species. Will we see the countryside with Bullfinches in most of the hedges, Song Thrushes in most gardens and Corn Buntings back in their old haunts? It is certainly possible but I fear I would not like to place any bets!

Redpath, S.M. & Thirgood, S.J. 1997 *Birds of prey and red grouse.* London: Stationery Office.
St. John, C. 1982 *A Scottish Naturalist: the Sketches and Notes of Charles St. John 1809-1856.* Andre Deutsch.

Common Sandpiper.

Using the Species Accounts

The heart of the book follows on pages 122-277. It gives the history of every breeding bird in Britain and Ireland over the 20th century, and an idea of what has happened to its populations, its importance in the European context and an assessment of what might happen as we go into the next millennium.

The species sections vary from a line or two for species which breed sporadically - or have held territory but have not, as yet, bred - to several hundred words for species for which we have a lot of information. The text can be easily understood without knowing exactly what the symbols and abbreviations stand for but these are there for readers who want to know a bit more about the birds involved. For more than 200 species where most (if not all) of these forms of data are available they appear at the top of the entry. All the birds appear in normal systematic order, whether they are obscure really wild birds or birds which are still mostly considered as escapes.

Name of species

Each bird has its English name and the scientific name. The English names used are the old style ones that have been used with little change for the last 40 years. New names calculated to allow easy reference to all the almost 10,000 bird species in the World have been ignored for the text of this book but, where they are different, appear in brackets after the English name at the head of the species entry. The scientific name of each species appears in the heading and, for a very few, subspecific names in italic are mentioned in the species texts.

Symbols

On the second line of each species text are up to four sets of symbols. They indicate the current status of the species in three official listings and my own assessment of the immediate prospects of the species:

these indicate the status of the species in the *Birds of Conservation Concern in the United Kingdom, Channel Islands and Isle of Man*. This was put together by the RSPB, BirdLife International, Wildfowl and Wetlands Trust, Game Conservancy Trust, BTO, Hawk and Owl Trust, the Wildlife Trusts and the National Trust in 1997. For a few species the listing does not relate to the bird's breeding status in Britain. Many species are not listed since the idea was to highlight the birds that were known to be doing badly. One and 2 symbols indicate amber and red listings respectively.

these indicate that the species will be on the Irish equivalent of the *Birds of Conservation Concern* when it is announced early in 2000. The shamrocks indicate listing in either Northern Ireland, or the Republic, or both. (1=amber;2=red list)

Gannets indicate the current state of the species if it is one assessed by the BTO for its alert lists. They reflect the BTO's conservation assessment of the species at the end of 1999:

indicates HIGH alert,

MEDIUM alert,

no problem at the moment and

a species RECOVERING.

Finally there are up to two smiley/grim/blank faces giving my subjective assessment of whether the species is going to do better or worse as it enters the 21st Century.

Boxed information

There is no way of presenting the numerical data within the body of the species texts without the writing getting very scrappy and boring. We know more about the distribution and populations of the birds in Britain and Ireland than any other comparable area in the World. The statistics given convey a good idea of the range of information available from a huge amount of work by a veritable army of volunteers. Exactly how these projects work and, most important, how you can help with them are given in the next chapter. The figures given are as follows:

DISTRIBUTION The two figures and percentages are from the Breeding Bird Atlases and relate to the 10-km squares where the birds were recorded. These are the number of squares with the bird in Britain and in Ireland for the 1988-1991 Atlas and the percentage change these figures represent from the 1968-1972 survey. The Isle of Man is included in Britain and the 14 squares in the Channel Islands are excluded.

NUMBERS BREEDING The estimates of the population (breeding pairs unless otherwise indicated) in both Britain and Ireland are given. Many of the figures for Britain are from the report in *British Birds* in January 1997 (Stone *et al.* 1997). Species marked RBBP are based on the data from the Rare Breeding Birds Panel up to 1997. Irish figures are generally based on the 1988-1991 Atlas figures. Where specific publications, apart from these, have been used to produce these estimates the source is referred to at the end of the species text. For a few common species, covered by the CBC or WBS, the British figures are modified by a + or - where there have been recent changes of more than about 20%, and ++ or – – if the changes are 50% more or less. A single figure is given for each species even though a range may have been shown in the original publication.

EUROPEAN STATUS These are the population figures given in the EBCC Atlas of European Breeding Birds for Europe without Russia and Turkey. The mid-point of the population estimate for the whole of Europe is given. This is the best estimate we have but obviously the figures are not as accurate from some areas as they are for others. For some species they are simply 'best guesses'. The percentage of the European population breeding in Britain and Ireland is then given using the figures available at the time the EBCC Atlas was compiled, and then, if this puts Britain and Ireland in the top ten of countries, what place we occupy.

BRITISH POPULATION TREND This is normally taken from *Breeding Birds in the Wider Countryside* (Crick *et al.* 1998) for the common species where a CBC (Common Birds Census) or WBS (Waterways Birds Survey) figure is available. These will generally be the CBC (all habitats) figure for 1972-96. If WBS data are given the interval is 1974-96. Other sources used include the CES (Constant Effort Sites) and BBS (Breeding bird Survey). These investigations are described in more detail in Investigating Our Breeding Birds. Just as the proofs of this book were being checked the BTO and RSPB published *The State of the UK's Birds 1999*. This includes populations trends of many species from 1970-98 from the CBC (or WBS) and these appear in {curly brackets} where they are available.

HOW LIKELY ARE YOU TO RECORD IT? The average number of 1-km squares used for the Breeding Bird Survey in Britain and Northern

Ireland from which the species was recorded over 1997 and 1998. The area surveyed in the two years was 4,491 1-km squares, which exceeded that of a large county like Hertfordshire in each year. For many species this gives a very good idea of just how easy - or difficult - it is to find the different species during the breeding season. Clearly it is a figure that represents a combination of ease of finding with area of distribution. It is NOT a good statistic to use for colonial breeding seabirds or other very localised birds. The percentage of squares (out of the 4,491) is given in brackets and the ranking out of all species is also given(1 being the most likely). This line is omitted for birds not recorded at all on the BBS in 1997 or 1998. In the Republic of Ireland the equivalent investigation is the CBS which covered 220 1-km squares in its first year (1998) - the ranking, from 1998 and 1999 combined, for these is given for the Republic of Ireland on the last line of the boxes in [brackets].

TEXT This is a discussion of what has happened and is happening to the species. These will be extended for species with a lot of information, shorter for those for which the information is less plentiful. These will generally have been discussed with the experts on each species. Each will end with my own personal guess, *in italics*, as to the prospects for each species over the next few years.

REFERENCES The general references (pages 283) include the major texts used in many different species' treatments, such as the Breeding Bird Atlases, and describes how they have been used. Specific papers published recently used, or which may be useful, for the species are listed with the author, year of publication, journal names, volume number and pages covered. For those species with Biodiversity Action Plans in the UK (UKBAP) a single line is also given listing the contact point (bold) and the lead partner or partners.

ILLUSTRATIONS These are by Kevin Baker and may be half page width or whole page and may include adjacent species - if they do they will be captioned, if not it will be obvious what the birds are from their placement on the page.

Hawfinch.

THE SPECIES

RED-THROATED DIVER

Gavia stellata

> *Distribution* Britain 379 (+22.3%) Ireland 10 (+42.9%)
> *Numbers breeding*: Britain 935 RBBP Ireland <10
> *European status*: 8,500 (16% in Britain and Ireland =3)
> *British population trend*: healthy and increasing
> *How likely are you to record it?* 32 squares (0.7%) *Ranked 125=*

These birds breed on tiny lochans and feed on the sea or adjacent larger waters mostly north and west of a line from Glasgow to Moray. Numbers reduced 100 years ago by persecution and collecting but continuing increase and range expansion in Scotland with survey estimate of 935 pairs in 1994. Irish population apparently less than 10 pairs but stable. Breeding failures at the young stage have increased recently and are worrying but Shetland productivity is very weather dependant. *Prospects seem good for further range expansion and filling in of gaps provided global warming does not degrade the conditions the birds require.*
Gibbons, D.W. *et al.* 1997 *Bird Study*: **44**, 194-205.

BLACK-THROATED DIVER

Gavia arctica

> *Distribution* Britain 199 (-6.1%) Ireland 0
> *Numbers breeding*: Britain 170 Ireland 1 ?
> *European status*: 22,000 (1% in Britain and Ireland =4)
> *British population trend*: seems assured
> *How likely are you to record it?* 8 squares (0.2%) *Ranked 155=*

This species usually nests and feeds on large freshwater lochs. There has been no marked change in Scotland save for a slow spread south but numbers have remained much the same with 151 summering pairs estimated in a 1985 survey which was the most complete count. The birds are absent from the Northern Isles and most of the Inner Hebrides but are on the Western Isles. A single possible breeding record in Donegal is recent and the first possibility recorded for breeding in Ireland. *The medium-term future seems assured for this magnificent Scottish bird.*

GREAT NORTHERN DIVER

Gavia immer

Only good record was a pair with young seen in Highland Region, Scotland, in 1997. Previous breeding record in 1970, which was possibly successful, and 1986, which failed, were certainly mixed pairs with Black-throated Diver.

PIED-BILLED GREBE *Podylimbus podiceps*

A hybrid pair of this American vagrant with a local Little Grebe raised three young in Cornwall during 1994.

LITTLE GREBE

Tachybaptus ruficollis

> *Distribution* Britain 1,275 (-6.6%) Ireland 336 (-34.4%)
> *Numbers breeding*: Britain 7,500 Ireland 4,500
> *European status*: 80,000 (15% in Britain and Ireland =1)
> *British population trend:* increasing (+83% WBS)
> *How likely are you to record it?* 100 squares (2.2%) *Ranked* **104** [93=]

These birds live in areas with thick vegetation growing in water – flowing or static. They are at risk from cold winters when the water may freeze. It is clear they have occupied much of Scotland for the first time in the

last few decades and some are now found in northern Caithness and on Orkney, but not Shetland. The trend from the WBS is not significant and the recorded loss, from the Atlas data, from Ireland is worrying. Global warming may be a good thing for these birds but the spread of the feral mink is a real worry. *On balance the next few years may be good for the Little Grebe.*

GREAT CRESTED GREBE

Podiceps cristatus

> *Distribution* Britain 892 (+17.1%) Ireland 225 (+0.9%)
> *Numbers breeding*: Britain 8,000 Ireland 4,150 (adults)
> *European status*: 290,000 (2% in Britain and Ireland)
> *British population trend*: increasing – long-term (+15% BBS)
> *How likely are you to record it?* 124 squares (2.8%) *Ranked* 100 [77=]

These birds are elegant and welcome breeders on open water through much of Britain south of the Highland fringe and about half of Ireland. In the mid 1800s they were severely depleted as they were taken for their skins – as substitutes for fur! The recovery was well underway by the turn of the century and they are still expanding. In Ireland Lough Neagh with 750 pairs is a very important site and they are beginning to colonise the south-eastern part of the country – which does not have as many suitable waters. In Scotland there may have been a very gradual spread northwards but this may have stopped about 30 years ago. *Good protection and new gravel pits are excellent news for this species.*

RED-NECKED GREBE

Podiceps grisegena

No successful breeding yet but up to ten scattered sites with birds in the summer each year in recent decades – mostly in England but most persistent is a pair nest building and incubating in Scotland. *Success any time now!*

SLAVONIAN GREBE

Podiceps auritus

> *Distribution* Britain 24 (+26.3%) Ireland 0
> *Numbers breeding*: Britain 56 RBBP Ireland 0
> *European status*: 7,500 (1% in Britain and Ireland =7)
> *British population trend*: fluctuating

Definite breeding was first proved in Scotland in 1908 and up to 80 pairs were breeding by the late 1970s and again in 1990 but there was a recent low of 37 in 1988 and there has been a worrying downward trend recently. Some breed away from core area but breeding productivity often very low. *Probably safe as a Scottish breeding bird.*

BLACK-NECKED GREBE

Podiceps nigricollis

> *Distribution* Britain 35 (+218%) Ireland 0
> *Numbers breeding*: Britain 46 RBBP Ireland 0
> *European status*: 33,000 (1% in Britain and Ireland)
> *British population trend*: apparently secure
> *How likely are you to record it?* 2 squares (0.0%) *Ranked* 183=

First proved breeding Anglesey 1904, Ireland 1915, England 1918 and Scotland 1930. Astonishingly a colony of 250 to 300 pairs was found in Co. Roscommon in 1929 but the lake was drained in 1934, and they have only been sporadic breeders in Ireland during the last 40 years. Up to 80 scattered pairs England and Scotland – mostly north of a line from Beachy Head to the Mersey – and possibly Anglesey in recent years. *Safer now they are dispersed?*

BLACK-BROWED ALBATROSS

Diomedea melanophris

One resided with the Gannets on Bass Rock in 1967 and 1968 and four times in 1969. It or another was at Hermaness, Shetland, definitely in 1972 and from 1974 to 1987 and 1990 to 1995. 'Albert' even built a nest in some years!

FULMAR (Northern Fulmar)

Fulmarus glacialis

> *Distribution* Britain 550 (+7.7%) Ireland 159 (+2.6%)
> *Numbers breeding*: Britain 539,000 Ireland 31,000
> *European status*: 2,800,000 (20% in Britain and Ireland =2)
> *British population trend*: still going up 3% or 4% p.a.
> *How likely are you to record it?* 46 squares (1.0%) *Ranked* 116= [84=]

The history of these gull-like tube-noses, over the last 100 years, is quite amazing and very well documented. At the turn of the century some 20,000 pairs bred on St Kilda and had been there for centuries. Other sites were colonised from Foula in 1878 followed by the isolated islands of the Outer Hebrides, Shetland, Orkney and even the mainland at Clo Mor in Sutherland. These colonists are thought to have originated from the Icelandic population and not St Kilda – where they were an important resource for the human population. The total outside St Kilda breeding in 1900 was in the low 100s and probably not as many as 500! Ireland was first colonised in Mayo in 1911, England (Yorkshire) in 1922 and Wales (Dyfed) in 1931. The current estimate for breeding pairs is 580,000 including 63,000 on the St Kilda group! We may confidently expect the figure to be higher for the current survey, Seabird 2000, since the population growth of at least 3-4% p.a. seems to be continuing.

These birds live seriously long lives and the oldest British ringing recovery is from Orkney at almost 41 years as a breeding bird. A few of the St Kilda birds may be over 100 years old! The huge increases are clearly powered by better survival of the birds at sea where they are able to, and do, exploit the discards from the fishing industry. However a genotype favouring expansion may also have appeared within the Icelandic population in the early nineteenth century. These birds have a very wide range of nest sites and nesting territories are not going to be limiting to the population. Further filling in of the lengths of coastline where there are possible nesting sites seems inevitable. *Prospects very rosy.*

MANX SHEARWATER

Puffinus puffinus

> *Distribution* Britain 22 (-31.3%) Ireland 13 (+18.8%)
> *Numbers breeding*: Britain 235,000 Ireland 40,000
> *European status*: 300,000 (92% in Britain and Ireland =1)
> *British population trend*: increasing – steadily?
> *How likely are you to record it?* 1 square (0.0%) *Ranked* 190=

These black and white shearwaters are normally only seen way out at sea from a ferry or flying past a headland. And yet our area has more than 90% of the world population! Most colonies are concentrated in rather small areas – the Kerry islands off Ireland, Skokholm and Skomer off Pembrokeshire and the island of Rum in the Inner Hebrides. The location of colonies has not changed much but it seems that there has been a big increase in numbers of birds at least on Skomer and Rum at each of which 100,000 or more pairs may breed! The birds are pretty invisible at the colonies as they come in under the cover of darkness but they are *very* noisy! As with the Fulmar the increase is probably due to better adult survival but *not* due to the exploitation of discards – they don't. The oldest Manx Shearwater was over 35 years.

The only English breeding sites are on the Scillies and Lundy – where there are a few left – and there are newly re-established breeding birds on the Calf of Man. Birds also breed on the Channel Islands. The infestation of islands by rats has caused the birds' local extinction in the

past and recent successful rat extermination techniques bode well for this and other burrowing species. Accurate counts at colonies are difficult but they are improving. These are very long distance migrants with the winter quarters off the coasts of Brazil, Argentina and Uruguay in South America. *A rosy future – without rats or mink!*

LITTLE SHEARWATER *Puffinus assimilis*

A male vagrant of the Madeiran race was territorial within the Manx Shearwater colony on Skomer (Wales) during the summers of 1981 and 1982.

STORM PETREL (European Storm-petrel)

Hydrobates pelagicus

> *Distribution* Britain 48 (+11.9%) Ireland 19 (+11.8%)
> *Numbers breeding*: Britain 80,000 ? Ireland 70,000 ?
> *European status*: 450,000 ? (33% in Britain and Ireland =2)
> *British population trend*: unknown – good?

This tiny black and white seabird – very similar in size to a House Martin but with all black plumage save for the white rump – is very difficult to spot in daylight at sea and only comes near land at night. It is not an easy bird to document and techniques for proper census of breeding colonies are only now being proposed and refined. Major colonies exist all along the West coast of Ireland, off Pembrokeshire and along the West coast of Scotland and in the Northern Isles. Where they have been lost, rats or cats are probably to blame. However there is a new factor as Great Skuas, which have only colonised St Kilda recently, are calculated to have taken

7,450 Storm Petrels in 1996! It is not at all certain what impact even this rate of predation might have on the breeding population. *Prospects good – were it not for the threat from Great Skuas.*

Gilbert, G., Hemsley, D. & Shepherd, M. 1998 *Scottish Birds*: **19**, 145-153.
Mainwood, A.R. *et al*. 1997 *Seabird*: **19**, 22-30.
Phillips, R.A., Thompson, D.R. & Hamer, K.C. 1999 *J. Appl. Ecol.*: **36**, 218-232.
Ratcliffe, N. *et al*. 1998 *Scottish Birds*: **19**, 154-159.
Wood, D. 1997 *Seabird*: **19**, 40-46.

LEACH'S PETREL (Leach's Storm-petrel)

 *Oceanodroma leucorhoa*

Distribution Britain 10 (+11.1%) Ireland 1 (0%)
Numbers breeding: Britain 50,000 ?? Ireland 500 ??
European status: 140,000 ?? (25% in Britain and Ireland =2)
British population trend: unknown – poor?

This petrel is a bit bigger than the Storm Petrel and even more enigmatic! Very few breeding sites are known and more may be undiscovered. Most colonies are on remote offshore Western Scottish islands though there are two small colonies in Ireland (Mayo and possibly Donegal). The Flannans may have several hundred – or thousand – pairs but St Kilda is the powerhouse with many thousands. Here Great Skua predation (see Storm Petrel above) is calculated to have killed 14,850 in 1996. This must be damaging as the birds have not had any equivalent adult predation to cope with in recent history. *Prospects dismal with almost all the bird's breeding at locations colonised by Great Skuas.*

Phillips, R.A., Thompson, D.R. & Hamer, K.C. 1999 *J. Appl. Ecol.*: **36**, 218-232.

SWINHOE'S PETREL (Swinhoe's Storm-petrel)

Oceanodroma monorhis

This browner cousin of the Leach's Petrel breeds in the western Pacific. However females have been trapped in Europe, including some with brood patches, and three have been caught on the coast of Tyne and Wear in the summer (1989 – 1993) – one individual in all of the last four of these years. *Who knows what they are doing!*

GANNET (Northern Gannet)

Morus bassanus

> *Distribution* Britain 18 (+28.6%) Ireland 5 (+66.7%)
> *Numbers breeding*: Britain 201,000 Ireland 30,000 nests
> *European status*: 230,000 (79% in Britain and Ireland =1)
> *British population trend*: on the up and up, recently by 2.4% p.a.
> *How likely are you to record it?* 10 squares (0.2%) *Ranked* 150 [88=]

The Gannet is the biggest, and most conspicuous, seabird native to the North Atlantic. Britain and Ireland hold about 80% of the European population. In 1900 there were nine colonies: one on Lundy, in the Bristol Channel, was about to become extinct but the rest are still extant. The total was 40,000 to 50,000 pairs. Now there were more than 230,000 pairs counted in 1994-95 in the following colonies:

Colony	County	Count	Year established
Bempton	Yorkshire	1,631	ca 1930
Bass Rock	Firth of Forth	34,397	(1447)
Troup Head	Grampian	530	1988
Fair Isle	Shetland	975	1974 or 75
Noss	Shetland	7,310	1914
Hermaness	Shetland	11,993	1917
Foula	Shetland	600	1976
Sule Stack	Orkney	4,888	(1710)
Sula Sgeir	Western Isles	10,440	(1549)
Flannan Isles	Western Isles	4,438	(1969)
St Kilda	Western Isles	60,428	(9th century)
Ailsa Craig	Firth of Clyde	32,456	(1526)
Scar Rocks	Wigtown	1,952	1939*
Clare Island	Mayo	3(1999)	1975 or 1976
Little Skellig	Kerry	26,850	(1700)
Bull Rock	Cork	1,815	1856
Great Saltee	Wexford	1,250	1929
Ireland's Eye	Dublin	ca. 150 (1999)	1989
Grassholm	Pembroke	29,277+	1820 or later
Ortac	Alderney	2.098	1940
Les Etacs	Alderney	3,380	(1945)

*first colonisation ca 1883 abortive

The population is still increasing at a rate of about 2.4% a year and so, by 1999, there are likely to have been about 260,000. The colony at Sula Sgeir is still harvested for young birds to eat by 'The Men of Ness' and it is not increasing. New colonies are likely to be established and these can take off very quickly to reach several hundred nests in a dozen years or so. Clearly these new colonies are occupied by birds bred elsewhere – survival and productivity are high and Gannets exploit fishing discards. One unpleasant threat comes from modern, indestructible, fish nets which entangle birds at sea and may also be incorporated into nests. *Excellent prospects – will they come back to nest on Lundy?*

Murray, S. & Wanless, S. 1996 *Scottish Birds*: **18**, 152-158.
Murray, S. & Wanless, S. 1997 *Scottish Birds*: **19**, 10-27.
Wanless, S., Matthews, J. & Bourne, W.R.P. 1996 *Scottish Birds*: **18**, 214-221.

CORMORANT (Great Cormorant)

Phalacrocorax carbo

Distribution Britain 174 (-20.0%) Ireland 93 (-4.1%)
Numbers breeding: Britain 7,000 Ireland 4,700
European status: 142,000 (8% in Britain and Ireland =4)
British population trend: stabilising? (+18% BBS)
How likely are you to record it? 296 squares (6.6%) *Ranked* 73 [63]

This large, black, fish-eating bird breeds in rather few colonies: mostly on islands, stacks and cliffs in remote coastal areas. The population, rather static, was severely persecuted by fishermen and most colonies were small (less than 200 pairs). Full protection was afforded in Britain in 1967 and in Ireland in 1976 and they increased markedly in the next ten years – Lambay Island (Dublin) held 1,027 pairs in 1985. At the same time further inland nesting sites in the western half of the country were established. In North-west Scotland there have been recent decreases, a modest increase in Wales and bigger increase in South-west Scotland. Historically a few small colonies were established (and destroyed) inland in East Anglia but the biggest change has been the recent colonisation, by tree nesting birds, of South-east England. Several colonies are now well established notably at Abberton (Essex) and Little Paxton (Cambs). These birds are mainly of the smaller race *sinensis* whose colonies on the Continent may reach several thousand pairs. Cormorants are still persecuted, mainly illegally, and research is underway to determine whether they are harmful to properly managed fisheries. *Current prospects very good – unless the legal goal posts are moved.*

Smiddy, P. 1998 *Irish Birds*: **6**, 213-216.

Cormorants.

SHAG (European Shag)

Phalacrocorax aristotelis

> *Distribution* Britain 386 (-5.5%) Ireland 123 (-14.0%)
> *Numbers breeding*: Britain 37,500 Ireland 8,800
> *European status*: 86,500 (55% in Britain and Ireland =1)
> *British population trend:* fluctuating
> *How likely are you to record it?* 15 squares (0.3%) *Ranked* 140=

This smaller cousin of the Cormorant is very much more a seabird and another colonial sea-cliff nester. It was widely persecuted as a fish-eater and for sport in the 19th century and increased this century as the killing diminished. The birds colonised the East Coast after they were virtually extinguished from England and became uncommon south of Caithness. And there may now be as many as 5,000 pairs along the North Sea coast of mainland Britain in a good year. It has been calculated that this increase averages about 10% per year for most of this century. However all is not well all the time as the population is liable to rather violent fluctuations and these may be becoming more frequent. One of these, affecting the Farne Islands (Northumbria) birds some 30 years ago, was tied to toxins produced by plankton – a so-called 'red tide' – but others are of unknown origin. For instance a very large proportion of the Isle of May breeding birds went missing over the winter of 1998/99. These events seem to be mostly restricted to the North Sea populations so their overall seriousness, for Shags over the whole of Britain and Ireland, may not be so important. *Population fluctuations cast a shadow over this species' future prospects.*

BITTERN (Great Bittern)

Botaurus stellaris

Distribution Britain 13 (-62.9%) Ireland 0
Numbers breeding: Britain 19 males Ireland 0
European status: 10,500 (0% in Britain and Ireland)
British population trend: falling but increase in 1999

This big striped brown heron is famous for its booming call – which enables research scientists to distinguish males as individuals from recordings. Nesting in Ireland and Scotland had ceased by 1840 and at the turn of the century the Bittern was lost as a breeding bird in England. After a few years they returned and more than 80 booming males were logged annually in the mid-1950s. There were birds in large reed-beds in several areas, including Anglesey and they were heard in a few places in Ireland. But by 1976 the population had dropped below 50 and a few years later none were breeding outside East Anglia and Lancashire. Much money and effort is being spent to retain these interesting and elusive birds. The total of at least 19 booming males in 1999 was encouraging after only 11 and 13 in previous two years. *Prospects good unless there are cold winters and provided the extensive work on habitat provision is kept up.* UKBAP **English Nature** RSPB.

LITTLE BITTERN   *Ixobrychus minutus*

This is a very difficult species to prove breeding and there are sporadic records, indicative of territory holding at the least, since about 1800. East Anglia and Kent are the usual, but not only, sites. Only instance of proved breeding was 1984 in Yorkshire (and present following year). Ogilvie, M.A. & RBBP 1999 *British Birds*: 92, 176-182 & 472-476.

NIGHT HERON (Black-crowned Night Heron)

Nycticorax nycticorax

A free flying colony of the American race (*hoactli*) has existed at Edinburgh Zoo since 1951 – five to ten birds currently breed. Also, at Great Witchingham in Norfolk, a similar sized colony of free-flying birds of the European race breeds in a Grey Heronry.

LITTLE EGRET

Egretta garzetta

Distribution Britain 0 Ireland 0
Numbers breeding: Britain 6 Ireland 12
European status: 28,000 (0% in Britain and Ireland)
British population trend: new colonist – outlook excellent
How likely are you to record it? 3 squares (0.1%) *Ranked* 171=

This elegant, small white heron has colonised Britain and Ireland from the south over the last few years. From 1989 the number of birds recorded in southern England increased apace and some played at breeding in existing Grey heron colonies from 1993. In 1996 breeding was recorded in Dorset (one pair) and there were 5 pairs there in 1997 with breeding (one or two pairs) in Hampshire. In southern Ireland no fewer than 12 pairs bred in a tree nesting Grey Heron colony. These birds seem only to be at risk if there is severe winter weather. *Hundreds in dozens of colonies by 2010!*
Lock, L. & Cook, K. 1998 *British Birds*: **91**, 271-272.
Smiddy, P. & O'Sullivan, O. 1998 *Irish Birds*: 6, 201-206.

GREY HERON

Ardea cinerea

Distribution Britain 2335 (+38.4%) Ireland 791 (+2.6%)
Numbers breeding: Britain 10,000 Ireland 3,650 nests
European status: 122,000 (11% in Britain and Ireland =3)
British population trend: increase 33% (sample census 1972-96)
How likely are you to record it? 989 squares (22.0%) *Ranked* 48 [38]

When falconry was important so were heronries: flying Peregrines at Herons was considered the very best sport but, subsequently, preservation for this purpose lapsed and the birds were depleted. There may have been about 6,000 breeding in England and Wales in 1890. Severe winter weather can cause big drops in breeding numbers – as much as 50% – but recovery to previous levels only takes a few years with mild winters. Currently, following a run of fairly open winters, the breeding populations may be around 3,000 pairs in Scotland, 750 in Wales and about 6,000 in England. The numbers of nests in England and Wales have been estimated at sample colonies for many years. Over the last few years the total for England has probably been at its highest recorded level – 6,000 to 6,300. In recent years illegal persecution from fishermen has probably declined and the effect of toxic chemicals, on a top predator, have diminished but deaths against wires continue. *Prospects good for a widespread species.*

WHITE STORK   *Ciconia ciconia*

A pair nested on St Giles Cathedral in Edinburgh in 1416. One or two birds have sometimes been present in East Anglia and other parts of England during the summer. An apparently free-flying pair in a North Nottinghamshire collection apparently bred successfully in 1999!

SPOONBILL (Eurasian Spoonbill)

 Platalea leucorodia

This species still bred in East Anglia until about 1650 having been lost from other English areas and Wales (Pembrokeshire) in 1602. Now up to 30 or more are present in summer and nest building recorded in East Anglia and Cheshire. *Any time soon!*

MUTE SWAN

Cygnus olor

> *Distribution* Britain 1,579 (-2.6%) Ireland 560 (-11.7%)
> *Numbers breeding*: Britain 25,750 Ireland 19,500 (adults)
> *European status*: 48,000 (48% in Britain and Ireland =1)
> *British population trend*: increasing still? (+59% WBS)
> *How likely are you to record it?* 363 squares (8.1%) *Ranked* 68 [64=]

These birds were an important status symbol in earlier times with bill-marks denoting the ownership. This was dying out 250 years ago but the birds were still considered as tame until earlier this century. They have increased gradually and may still be increasing with better protection and more wetlands for them to use. Lead poisoning was an important mortality factor in the 1960s until the material was banned from being used by anglers (Britain 1987) and many populations started to increase immediately. The ban on lead loaded shotgun use over water on certain areas (1999 England and Wales) may help further. Current breeding population estimates might be of the order of 200 pairs in Wales, 550 in Scotland, over 4,000 in Ireland and almost 6,000 in England. However the translation of number of adults to breeding pairs is always tricky. Colonial breeding occurs at the Fleet in Dorset, Loch of Harray in Orkney and in Ireland in Donegal and Wexford. *This species has a special place in public affection in many areas and prospects are good.*

BLACK SWAN *Cygnus atratus*

Several dozen escaped birds (the species comes from Australia) reported in the last decade. Breeding reported in Northamptonshire, Lothian, Essex and Wiltshire 1996 or 1997.
Ogilvie, M.A. & RBBP 1999 *British Birds*: **92**, 176-182 & 472-476.

TRUMPETER SWAN *Cygnus buccinator*

First breeding of this North American species, never recorded wild in Europe, reported near local wildfowl collection in Northamptonshire in 1997.
Ogilvie, M.A. & RBBP 1999 *British Birds*: **92**, 176-182 & 472-476.

WHOOPER SWAN

Cygnus cygnus

Occasional wild pairs breed in Scotland and even Ireland now, but a colony regularly breeding on Orkney stopped as early as about 1770. Mixed pairs with Mute Swans sometimes reported and up to two dozen feral pairs reported too. Overall less than 5 pairs successful in any year.
Ogilvie, M.A. & RBBP 1999 *British Birds*: **92**, 176-182 & 472-476.

PINK-FOOTED GOOSE

Anser brachyrhynchus

Many summer records in Scotland (and England) are of 'pricked' birds – injured by wildfowlers – and 69 adults and 7 juveniles counted in summer 1991. The only breeding pair found was in Lancashire. Mixed pairs, often with Greylags, are sometimes reported.
Delany, S. 1993 *British Birds*: **86**, 591-599.
Ogilvie, M.A. & RBBP 1999 *British Birds*: **92**, 176-182 & 472-476.

WHITE-FRONTED GOOSE

(Greater White-fronted Goose)

Anser albifrons

In the last ten years up to four pairs in Argyll and one in Greater Manchester of the Greenland race (*flavirostris*) have bred and one, of the nominate race in Norfolk, was recorded in the 1991 survey which found a total of 77 birds of this species.

Delany, S. 1993 *British Birds*: **86**, 591-599.

Ogilvie, M.A. & RBBP 1999 *British Birds*: **92**, 176-182 & 472-476.

GREYLAG GOOSE

Anser anser

> *Distribution* Britain 718 (+258%) Ireland 23 (+188%)
> *Numbers breeding*: Britain 14,300 Ireland 700 adults
> *European status*: 55,000 (20% in Britain and Ireland =2)
> *British population trend*: increasing at about 5% p.a. (+31% BBS)
> *How likely are you to record it?* 183 squares (4.1%) *Ranked* 91= [101=]

These birds are real wild members of the British and Irish avifauna but the wild birds have long been confined to the Western Isles and northern Scotland. Here the birds are increasing and may number almost 1,000 pairs. Elsewhere re-introduced Greylag Geese and feral populations have established themselves in many areas – particularly eastern England, Shropshire, the Lake District, Anglesey, South-west Scotland, the Lowlands and Strangford Loch and Co. Wexford. In Ireland they do not seem to be increasing much but they are expanding range and increasing rapidly in Britain – probably 5% or more per year. There are many areas where this species is increasing much faster than the Canada Geese. Hybrid breeding with other species is quite frequently reported and there are areas where they cause problems with other species. *Increase is set to carry on – the return of the native after over 200 years!*

Delany, S. 1993 *British Birds*: **86**, 591-599.

BAR-HEADED GOOSE *Anser indicus*

These are Asian geese and 85 were recorded in the 1991 survey. Breeding reported in 1996 and/or 1997 from Hampshire, Derbyshire, Nottinghamshire and Greater Manchester.

Delany, S. 1993 *British Birds*: **86**, 591-599.

Ogilvie, M.A. & RBBP 1999 *British Birds*: **92**, 176-182 & 472-476.

SNOW GOOSE *Anser caerulescens*

There are several races of these North American geese represented in the feral population in Britain. They were recorded at 27 sites (182 birds) in 1991. The population of 40 to 50 birds centred on Coll and Mull regularly breed. Other recent breeding records come from Hampshire, Berkshire, Norfolk and Bedfordshire.

Delany, S. 1993 *British Birds*: **86**, 591-599.
Ogilvie, M.A. & RBBP 1999 *British Birds*: **92**, 176-182 & 472-476.

CANADA GOOSE

Branta canadensis

Distribution Britain 1,196 (+75.6%) Ireland 19 (+217%)
Numbers breeding: Britain 46,700 Ireland 650 adults
European status: 40,000 (76% in Britain and Ireland =1)
British population trend: increasing by 2.4% p.a. and slowing
How likely are you to record it? 620 squares (13.8%) *Ranked* 57 [93=]

This is an introduced species from North America first brought over sometime earlier than 1678 when they were recorded in St James's Park, London. For about 250 years few were present outside the country seats where gentleman chose to have ornamental wildfowl collections. Then they started to become more common as escapes and the runaway increase began. Now most of England is colonised and there are areas in the other countries where they are well established. In Ireland, where the annual increase rate may be about 3%, most of the 500 adult birds are on the Woodford River (Cavan/Fermanagh). The English population was increasing at a calculated rate of 9.8% before 1985/86 but, more recently it has dropped back

to 2.4%. The British population (rather few in Scotland and Wales away from the borders and North coast) may now be over 56,000 adult birds. These birds are considered to be a real nuisance as they will form large, and dense, breeding concentrations in very close proximity to man. These may be in country house parks, gravel pits, reservoirs or even suburban and urban lakes. Often they are in situations where sport shooting is not appropriate and control by pricking eggs and rounding up moulting birds has been proposed. Much research is in progress at the moment. Ringing has shown that there is a well developed moult migration from the North Midlands to Beauly Firth and that some birds go south to France in cold winters. *Prospects good, if one wants them, but population may be reduced by control.*

Browne, A.M. & O'Halloran, J. 1998 *Irish Birds*: 6, 233-236.

Delany, S. 1993 *British Birds*: 86, 591-599.

DETR (various authors). 1998 **Population Dynamics of Canada Geese.**

BARNACLE GOOSE

Branta leucopsis

Not yet considered as a breeding bird by many people but there are already feral populations breeding in several places in England and around Islay in Scotland and Strangford Lough in Ireland. About 730 adults in Britain and 80 in Ireland.

Delany, S. 1993 *British Birds*: 86, 591-599.

Ogilvie, M.A. & RBBP 1999 *British Birds*: 92, 176-182 & 472-476.

EGYPTIAN GOOSE

Alopochen aegyptiacus

> *Distribution* Britain 87 (+383%) Ireland 0 (0%)
> *Numbers breeding*: Britain 700 adults Ireland 0
> *European status*: 1,750 (22% in Britain and Ireland =2)
> *British population trend*: increasing and expanding
> *How likely are you to record it?* 3 squares (0.1%) *Ranked* 171=

These strangely coloured geese were brought over from Africa to grace ornamental collections over 200 years ago. Up to 40 years ago there was only a small breeding population of feral birds in Norfolk. Now they breed through the county and into Suffolk with nests in recent years in Buckinghamshire, Hampshire, Rutland, Surrey, Berkshire, Essex and

Greater Manchester. There might be as many as 2,000 adults present in England now. They start to breed early and take over large cavities in trees and deprive other species, like Kestrels and Barn Owls, of their traditional nest sites. *Probably unstoppable.*
Delany, S. 1993 *British Birds*: 86, 591-599.
Ogilvie, M.A. & RBBP 1999 *British Birds*: 92, 176-182 & 472-476.

RUDDY SHELDUCK *Tadorna ferruginea*

These birds are fairly often seen in England and sometimes in Scotland. Recently breeding by one pair has been reported from Norfolk and one paired (successfully) with an Egyptian Goose (female) at Rutland Water in 1996 after five years unproductive life together.
Ogilvie, M.A. & RBBP 1999 *British Birds*: 92, 176-182 & 472-476.

CAPE SHELDUCK (South African Shelduck)

Tadorna cana

A pair with four juveniles in Surrey in 1997 would be the first breeding in the wild in Britain – but were they unconfined when they nested?
Ogilvie, M.A. & RBBP 1999 *British Birds*: 92, 176-182 & 472-476.

SHELDUCK

 Tadorna tadorna

Distribution Britain 959 (+19.7%) Ireland 183 (-14.9%)
Numbers breeding: Britain 10,600 Ireland 1,100
European status: 41,000 (28% in Britain and Ireland =1)
British population trend: increasing, but down 35% on BBS
How likely are you to record it? 260 squares (5.8%) *Ranked* 80= [71=]

This is a striking bird, not quite a duck and not quite a goose, that looks black and white at a distance. At the turn of the century they were beginning to recover. They had earlier been eradicated from some areas, for example Breckland, because they competed for their burrows with valuable rabbit stocks! In other parts of South-east England they had been shot for sport. Now these birds are increasing and nest in most coastal areas with substantial populations away from the sea. Many people expect these to be only a mile or two inland but many birds now nest well inland especially in East Anglia, up the Trent and other parts of the Midlands, in the New Forest and in lowland Scotland. There is no evidence of big population change in Ireland where there may have been some contraction of range and birds do not often breed inland. In Britain the major increase has been in England and Wales (over 1,000 pairs) where more and more birds are colonising inland sites. The Scottish population has been estimated at about 1,500 pairs. These birds have major moulting areas where they congregate in the autumn and a system of crèches so that a few 'aunties' are left in charge of large numbers of ducklings, often of different sizes. *Prospects look good but feral mink are likely to be a problem for the nesting birds.*

MUSCOVY DUCK *Cairina moschata*

These birds, originally from South America, are often ignored by birdwatchers. They are present and breeding in several places with flocks of more than 100 reported. Breeding recorded in 1996/97 from Devon, Northumberland, Surrey, Bedfordshire, Cambridgeshire, Derbyshire, Norfolk and Nottinghamshire (and probably on a pond near you!).
Ogilvie, M.A. & RBBP 1999 *British Birds*: **92**, 176-182 & 472-476.

WOOD DUCK *Aix sponsa*

Recorded in 30 squares (proved in 6) in latest Atlas and 24 pairs found in Kent during 1996 where there may be two or three hundred of this North American relative of the Mandarin currently at large.
Ogilvie, M.A. & RBBP 1999 *British Birds*: **92**, 176-182 & 472-476.

MANDARIN

Aix   galericulata

> *Distribution* Britain 218 (+459%) Ireland 0
> *Numbers breeding*: Britain 7,000 individuals Ireland 0
> *European status*: 5,000 ? (70% in Britain and Ireland =1)
> *British population trend*: increasing and expanding
> *How likely are you to record it?* 18 squares (0.4%) *Ranked* 138

These pretty tree-hole nesting little ducks have been kept in (and escaped from) collections in Britain for 250 years. Recent expansion and increases have led to an estimate of 7,000 individuals for Britain. This means that the population is about a tenth of the truly wild population in the Far East (mainly Japan). Most are in the Home Counties but they are present in several other English areas (Severn, Rutland and Norfolk), a few in Wales and in several places in Scotland (Perth, Borders, Argyll and Highland) and, in the last 10 years, in Co. Down. A *further increase in prospect.*
Ogilvie, M.A. & RBBP 1999 *British Birds*: **92**, 176-182 & 472-476.

WIGEON (Eurasian Wigeon)

Anas penelope

> *Distribution* Britain 360 (+27.2%) Ireland 25
> *Numbers breeding*: Britain 400 Ireland 0
> *European status*: 105,000 (0% in Britain and Ireland =5)
> *British population trend*: probably increasing
> *How likely are you to record it?* 14 squares (0.3%) *Ranked* 143[93=]

Although thought of as a winter visitor this is a not uncommon breeding duck. The birds in northern England and Scotland are probably of wild origin but some of the southern birds originate from releases and escapes. There may be 400 pairs in Scotland particularly Orkney and the Outer Hebrides. A few breed in Wales (mostly Anglesey) and a handful in Ireland – the others mainly breed in the Pennines and eastern Britain. *Outlook seems good for increase and spread.*

GADWALL

Anas strepera

> *Distribution* Britain 357 (+126%) Ireland 25 (+78.6%)
> *Numbers breeding*: Britain 770 Ireland 30
> *European status*: 23,000 (3% in Britain and Ireland =8)
> *British population trend*: excellent prospects
> *How likely are you to record it?* 50 squares (1.1%) *Ranked* 114

About 150 years ago pinioned wild-caught Gadwall were first introduced in Norfolk and the population there gradually built up. Wild birds (from the Icelandic population) may have bred in Scotland sometimes but the first birds to breed at Loch Leven arrived about 90 years ago and up to 40 pairs may nest there in some years. There are now populations in several parts of the Lowlands, Caithness and some parts of the Hebrides – probably well under 100 pairs. In Ireland breeding was first proved about 70 years ago and the increasing population may soon reach 50 pairs – mainly at Lough Neagh and in Galway and Wexford. In Wales a handful may breed regularly in Anglesey and there have been sporadic records in other areas. There may now be approaching 1,000 pairs in England. *Prospects excellent in many areas.*

TEAL (Common Teal)

Anas crecca

> *Distribution* Britain 1,147 (-16.6%) Ireland 188 (-51.5%)
> *Numbers breeding*: Britain 2,050 Ireland 550
> *European status*: 350,000 (1% in Britain and Ireland =10)
> *British population trend*: declining
> *How likely are you to record it?* 39 squares (0.9%) *Ranked* 122 [84=]

These small ducks are not easy to record when breeding, but they probably underwent a considerable decline in the 19th century as land was drained and enclosed. Their breeding range, and numbers, probably stabilised in most upland areas, particularly in Scotland, and some lowland. However there is good evidence for decline in population (and range) over the last 20 or 30 years with the effects of planting trees for forestry being partly responsible. Up to 150 pairs may now nest in Wales. In Ireland the south-eastern half of the island now has very few breeding records and the birds were found in less than half as many squares for the second as for the first breeding atlas. In winter the numbers of immigrant Teal dwarfs those from native stock. *Prospects do not look good.*

MALLARD

Anas platyrhynchos

Distribution Britain 2,596 (-1.2%) Ireland 831 (-9.1%)
Numbers breeding: Britain 115,000 Ireland 23,000
European status: 2,200,000 (6% in Britain and Ireland =7)
British population trend: increasing (+29% CBC, +65% farm){+82%}
How likely are you to record it? 1978 squares (44.0%) Ranked 30 [34]

The Mallard is the 'Wild Duck' and can be found breeding, even far from water, in all sorts of habitats from remote upland moors to city centre sixth-floor window boxes! It is clearly difficult – impossible – to tell birds originating from populations released by wildfowlers or 'tame' birds. The census work of BTO members indicates good increases following the cold winters of 1962 and 1963 – probably doing better than doubling before 1972. The WBS figures indicate an increase of 199% – almost three-fold – in the 23 years from 1974 to 1976 and the BBS shows no sign of any very recent decrease (+5% non-significant). The recently reported decline in wintering numbers is thus most likely to be due to a decline in winter visitors from the Continent. Most domesticated varieties of duck are derived from Mallard stock – the drakes show the characteristic curled feathers above the tail – and so many areas are blessed (or cursed) with rather strange birds. They may be bigger or smaller, darker, pure white or even sport top-knots! The drakes practise serial monogamy seeking a new mate as soon as their first duck has settled on her first clutch. This leads to distressing gang rapes of any single females where several drakes are gathered. *Mallards are set to continue their successful exploitation of man.*

BLACK DUCK (American Black Duck) Anas rubripes

Both a duck and a drake of this North American species have been present at different times in Scilly and the former bred several times with a Mallard drake.

PINTAIL (Northern Pintail)

Anas acuta

> *Distribution* Britain 85 (-1.2%) Ireland 9 (-25.0%)
> *Numbers breeding*: Britain 33 RBBP Ireland <1
> *European status*: 28,000 (0% in Britain and Ireland)
> *British population trend*: possibly stable
> *How likely are you to record it?* 3 squares (0.1%) *Ranked* 171=

This fine duck has very precise habitat requirements and is now, at best, a sporadic breeder over most of our area. This is an improvement on 100 years ago when Loch Leven (Kinross) was the only regular site. Now Orkney, Tiree (Argyll) and the washlands of the Fens have established populations and there are scattered records in all four nations. Some of these birds may have originated from feral populations and deliberate introductions. *Prospects not particularly good – or bad!*
Fox, A.D. & Meek, E.R. 1993 *British Birds*: **86**, 151-163.

GARGANEY

Anas querquedula

> *Distribution* Britain 138 (+1.5%) Ireland 8 (+60.0%)
> *Numbers breeding*: Britain 129 RBBP Ireland <1
> *European status*: 85,000 (0% in Britain and Ireland)
> *British population trend*: doing well
> *How likely are you to record it?* 1 square (0.0%) *Ranked* 190=

This small duck is a long distance migrant wintering in Africa. A hundred years ago it was virtually confined to East Anglia and Kent but the population increased and expanded – extremely sporadic in Ireland, now and again in Wales (mainly Anglesey) and recently regular with up to 20 pairs in Scotland. It is now fairly regular in many areas of England with a good number living on reserve areas. The best RBBP totals in recent years were maxima of 160 and 163 pairs in 1992 and 1993. *Prospects seem rather good.*

BLUE-WINGED TEAL *Anas discors*

A pair present in Essex from 1994 bred successfully in 1997. Britain and Ireland's first record of breeding for this migrant North American duck but, undoubtedly, escapes.
Ogilvie, M.A. & RBBP 1999 *British Birds*: **92**, 176-182 & 472-476.

SHOVELER (Northern Shoveler)

 ☺

Anas clypeata

> *Distribution* Britain 454 (-12.7%) Ireland 45 (-11.8%)
> *Numbers breeding*: Britain 1,250 Ireland <100
> *European status*: 37,000 (4% in Britain and Ireland =7)
> *British population trend*: probably stable
> *How likely are you to record it?* 21 squares (0.5%) *Ranked* 135

In the 19th century these striking ducks were in a parlous state with breeding only regular in East Anglia and south-east Scotland. Unregulated wildfowling was blamed and as legislation started to work these ducks, and other wildfowl, began to expand as breeding birds towards the end of the century. They may have bred earlier in Ireland but widespread breeding, by up to 100 pairs, might be a rather recent phenomenon. The best concentrations are at Loch Neagh and in the Shannon Callows. In Wales Anglesey is the only area with regular breeding. In Scotland breeding in Orkney, the Lowlands, part of the Hebrides, the Stewartry and the Borders may account for 100 or more pairs in a good year. This means that the English population may top 1,000 pairs in most years. The population may have peaked a decade ago but many nest in protected areas. *Increased interest in wetlands must be good news.*

RED-CRESTED POCHARD *Netta rufina*

Escaped birds sometimes bred in England from 60 years ago but recently records have increased greatly. Two or three dozen pairs now reported mainly round the Cotswold Water Park (Glos/Wilts) and near Pensthorpe (Norfolk).

Ogilvie, M.A. & RBBP 1999 *British Birds*: **92**, 176-182 & 472-476.

POCHARD (Common Pochard)

Aythya ferina

> *Distribution* Britain 511 (-0.2%) Ireland 40 (-21.6%)
> *Numbers breeding*: Britain 380 RBBP Ireland 30
> *European status*: 220,000 (0% in Britain and Ireland)
> *British population trend*: increasing
> *How likely are you to record it?* 30 squares (0.7%) *Ranked* 129=

These birds, the larger of the two widespread diving ducks, started breeding in England about 1800 and, by 100 years ago, had colonies in Norfolk, Hertfordshire, Essex and Yorkshire. There was scattered breeding elsewhere, mainly in eastern counties, and throughout Scotland. Here, in recent years, the Lowlands and Orkney are most regularly occupied but there are probably less than 50 pairs in the country. In Ireland they nested sporadically in the first 70 years of the century but, more recently, have increased to become regular at half a dozen or more sites. Up to 20 pairs may now breed in Wales – mostly Anglesey. In England the increase in water bodies from the building of reservoirs to the digging of gravel pits has been very good for these birds. The winter population is much inflated by birds from the extensive breeding grounds to our east through Europe and well into Asia. Non-breeding birds regularly summer but, even allowing for this, there could be as many as 600 pairs breeding in a good year. Precise figures are difficult to gather systematically for widespread and not uncommon species. *Set to carry on expanding.*

TUFTED DUCK

Aythya fuligula

> *Distribution* Britain 1,484 (+15.1%) Ireland 252 (-24.1%)
> *Numbers breeding*: Britain 7,500 Ireland 1,800
> *European status*: 280,000 (3% in Britain and Ireland =7)
> *British population trend*: expanding and increasing (+20% WBS)
> *How likely are you to record it?* 261 squares (5.8%) *Ranked* 79 [93=]

This species colonised Britain and Ireland in the second half of the 19th century – much in the last quarter. In fact most of southern and eastern Scotland was occupied over this period and breeding in Ireland was initiated at Lough Neagh. They continued to spread over the last 100 years and there are now many more suitable waters. The spread of the freshwater zebra mussel – released into London docks in 1824 – may have fuelled the bird's notable increase. It is certainly an important food item. In Ireland they now breed on many suitable waters, mostly above the line from Dundalk to Limerick, but have recently declined on Lough Neagh from 1,000 to 300 breeding pairs. The lack of suitable lakes, at low enough levels, may be responsible for their absence from the south-west peninsula, most of Wales and much of the Highlands. Over 1,000 pairs probably breed in Scotland – including about 400 at the remarkable colony on St Serfs Island in Loch Leven. The feral mink infestation is one of the few obvious problems for the future. Both WBS and BBS show gradual, but not significant increases in population. *Prospects good for increasing numbers.*

SCAUP (Greater Scaup)

Aythya marila

About 60 breeding attempts recorded mostly in the north of Scotland, particularly Orkney, and twice in Lincolnshire, once in Wales (1988), but never in Ireland until 1997 in the North. Otherwise no breeding records in our area since 1989.

EIDER (Common Eider)

Somateria mollissima

Distribution Britain 488 (+5.7%) Ireland 45 (+18.4%)
Numbers breeding: Britain 31,500 Ireland 800
European status: 850,000 (4% in Britain and Ireland =5)
British population trend: increasing at 2.5% p.a.(Atlas text)
How likely are you to record it? 12 squares (0.3%) *Ranked* 145=

These large coastal breeding ducks increased greatly during the 19th century spreading from the Scottish Islands to parts of the mainland and

a foothold in England in the North-east. First breeding in Ireland was in 1912 (Donegal) and they are now round the coasts of the northern quarter of the island, including Sligo. In Scotland virtually all coasts are colonised except the Solway and there are some notable concentrations – like the Sands of Forvie near Aberdeen. On the English west coast the thriving colony at Walney Island is virtually the only breeding site. *Gradual spread South expected.*

LONG-TAILED DUCK

Clangula hymealis

A few sporadic breeding attempts recorded in Orkney and Shetland up to 50 years ago. More recently may have bred in the Western Isles in 1969.

COMMON SCOTER (Black scoter)

Melanitta nigra

> *Distribution* Britain 51 (+21.4%) Ireland 16 (+6.7%)
> *Numbers breeding*: Britain 100 RBBP Ireland 65
> *European status*: 7,000 (2% in Britain and Ireland =5)
> *British population trend:* probably declining
> *How likely are you to record it?* 2 squares (0.0%) *Ranked* 183=

These seaduck breed on freshwater lakes and were in Northern Scotland in very small numbers 100 years ago. Now Caithness and Sutherland probably have more than half the breeding birds. In Ireland a colony at Lower Lough Erne, established 1905, peaked at 150 pairs (1967) and was extinct by 1995, when the three other Irish colonies held up to 110 pairs. *Eutrophication of the Irish lakes is a grave problem and feral mink predation a real threat.*

Gillings, T. & Delany, S. 1996 *Irish Birds*: 5, 413-422.
Underhill, M.C. *et al.* 1998 *Bird Study*: 45, 146-156.
UKBAP *Scottish Natural Heritage* RSPB & WWT.

VELVET SCOTER

Melanitta fusca

Breeding sometimes suspected, last time in Shetland in 1945, but never proved.

GOLDENEYE (Common Goldeneye)

Bucephala clangula

> *Distribution* Britain 173 (+811%) Ireland 13
> *Numbers breeding*: Britain 200 Ireland 0
> *European status*: 250,000 (0% in Britain and Ireland)
> *British population trend*: new and with excellent prospects
> *How likely are you to record it?* 5 squares (0.1%) *Ranked* 162=

Early Scottish breeding records of this hole-nesting species have all been dismissed but two injured birds nested in Cheshire in 1931 and 1932. Breeding, in specially erected nest boxes in Speyside, started in 1970 and has taken off. They now breed at several sites along the Spey and in adjacent areas where there are boxes. In Lancashire released birds are now trying to breed. The lack of nest sites, because we do not have the big Black Woodpeckers whose holes are large enough, is widely thought to have excluded them. *Prospects excellent – put up more boxes!*

RED-BREASTED MERGANSER

Mergus serrator

> *Distribution* Britain 674 (+1.5%) Ireland 167 (-32.7%)
> *Numbers breeding*: Britain 2,200 Ireland 700
> *European status*: 70,000 (4% in Britain and Ireland =5)
> *British population trend*: gradual expansion
> *How likely are you to record it?* 25 squares (0.6%) *Ranked* 132 [101=]

This species is the smaller of our two breeding sawbills and breeds both on the coast and along rivers and by lakes. By 1900 it was breeding through most of Scotland north of the Lowlands and over much of Ireland, save some of the south-east, but none bred in England or Wales. Birds started to breed in southern Scotland and then in England (in Cumbria 1959) and Wales (Anglesey 1953, mainland 1957). In Ireland it may have declined slightly recently and it is much persecuted, as in parts of Scotland, for perceived damage to salmonid fisheries. It now breeds all over the Lake District and in the Peak District (?100 in England) and North-west Wales (almost 200 pairs). *Probably good for further expansion.*

Gregory, R.D., Carter, S.P. & Baillie, S.R. 1997 Bird Study: 44, 1-12.

GOOSANDER

Mergus merganser

> *Distribution* Britain 674 (+64.0%) Ireland 2 (+100%)
> *Numbers breeding*: Britain 2,600 Ireland 1
> *European status*: 53,000 (5% in Britain and Ireland =3)
> *British population trend*: expanding and increasing
> *How likely are you to record it?* 78 squares (1.7%) *Ranked* 107

These birds started to colonise Scotland in about 1870 and were fairly common north of the Lowlands in mainland areas by 1900. They almost all breed inland on rivers and use holes for their nests. They are hated by many sport fishermen and persecuted (both legally and illegally) especially in Scotland. Nevertheless they have spread southwards and the first English breeding was Northumberland (1941) followed by colonisation of the uplands north of a line from the Flamborough to the Ribble. Now some in the Peak District and even Devon and Wales have been colonised over the last 30 years. A single pair has been recorded in North-west Ireland, on and off, for three decades, with breeding proved in Wicklow in 1997. *Further expansion to be confidently expected.*
Gregory, R.D., Carter, S.P. & Baillie, S.R. 1997 *Bird Study*: **44**, 1-12.

RUDDY DUCK

Oxyrura jamaicensis

> *Distribution* Britain 292 (+144%) Ireland 8 (0%)
> *Numbers breeding*: Britain 570 Ireland 20
> *European status*: 600+ (95% in Britain and Ireland =1)
> *British population trend*: further increase – unfortunately!
> *How likely are you to record it?* 16 squares (0.4%) *Ranked* 139

This attractive little American stiff-tail duck has, unfortunately, become well established in England after some escaped from Slimbridge in the late 1950s. By 1975 about 50 pairs bred (including Ireland's first – and Scotland in 1979). Breeding totals may now be 1,000 pairs in England and Wales, 50 in Ireland and possibly 100 in Scotland. This is a conservation disaster. Ruddy Ducks have reached the Spanish nesting area of the old-world stiff-tail, the White-headed Duck (*Oxyrura leucocephala*). Ruddy drakes are dominant so their presence hazards the endemic Old World species and *steps are to be taken* to eliminate the problem. *Hopefully their time will soon be up!*
Hughes, B., Underhill, M. & Delany, S. 1998 *British Birds*: **91**, 336-353.
Perry, K.W., Wells, J.H. & Smiddy, P. 1998 *Irish Birds*: **6**, 217-224.

HONEY BUZZARD (European Honey Buzzard)

 Pernis apivorus

> *Distribution* Britain 27 (+146%) Ireland 0
> *Numbers breeding*: Britain 55 Ireland 0
> *European status*: 45,000 (0% in Britain and Ireland)
> *British population trend*: steady increase
> *How likely are you to record it?* 1 square (0.0%) *Ranked* 190=

Always a rare migrant breeder, with many secret records, there has been a recent increase to about 50 or 55 pairs, mostly at established sites in English woodlands (but Atlas records also from Scotland). First breeding in Wales was in 1992. The population is creeping up from the sporadic breeding reported earlier. Better protection and the possible increase of prey (wasps and bees) through global warming may be helping. *The British population may gradually reach about 200 pairs.*
Roberts, S.J., Lewis, J.M.S. & Williams, I.T. 1999 *British Birds*: **92**, 326-345.

RED KITE

 *Milvus milvus*

> *Distribution* Britain 85 (+150%) Ireland 0
> *Numbers breeding*: Britain 300 Ireland 0
> *European status*: 22,000 (0% in Britain and Ireland =9)
> *British population trend:* increasing – possibly 30% p.a.
> *How likely are you to record it?* 27 squares (0.6%) *Ranked* 131

Few birds have teetered on the brink of extinction and then come back with such success. These birds were once (up to the 18th century) widespread but the population was persecuted and languished to the point where a single breeding female was successfully breeding in the 1930s in Central Wales. There followed a gradual increase and expansion on to better ground, probably helped by the arrival of new genes about 30 years ago. This enabled the breeding success to improve and the population increased apace. In spring 1999 there were 228 pairs and 155 unpaired birds and at least 165 young were reared. By itself this could be the success story of the 20th century: but there is more! Introductions of Spanish birds to England and Swedish birds to Scotland from 1989, and then other areas, proved an outstanding success for the RSPB, EN and SNH. Further sites are now involved and 116 pairs nested producing about 200 young in 1999. Increases of 30% per year are now

logged with the new birds settling close to the release sites. The only hint of a setback comes from the highly toxic rodenticides introduced because of the resistance of rats to warfarin. Red Kites may scavenge the dead bodies and pick up the poison, which has been found in several bodies of dead birds found recently. *Outlook brilliant: possibly 2,500 pairs by 2010 if illegal (and accidental) poisoning is curbed.*
Davis, P.E. 1993 *British Birds*: **86**, 295-299.
Dennis, R. H. *et al.* 1997 *British Birds*: **90**, 123-138.

WHITE-TAILED EAGLE

 *Haliaeetus albicilla*

> *Distribution* Britain 9 (0%) Ireland 0
> *Numbers breeding*: Britain 11 RBBP Ireland 0
> *European status*: 2,400 (1% in Britain and Ireland)
> *British population trend*: very gradual increase possible
> *How likely are you to record it?* 2 squares (0.0%) *Ranked* 183=

Rendered extinct in 1917 Scotland (?1898 Ireland and 1794 England). Successful reintroduction of Norwegian birds started on Rum by NCC (now run by SNH & RSPB) in 1975 and first chick was reared by a wild pair in 1985. Progress currently very slow but sure, but some have been illegally killed. *It is planned that more birds will be released regularly for a truly viable population to be established.*

MARSH HARRIER (European Marsh Harrier)

Circus aeruginosus

> *Distribution* Britain 114 (+339%) Ireland 7 (+600%)
> *Numbers breeding*: Britain 131 RBBP Ireland 0
> *European status*: 29,000 (0% in Britain and Ireland)
> *British population trend*: prospects excellent
> *How likely are you to record it?* 20 squares (0.2%) *Ranked* 136=

Young birdwatchers may find it amazing that this species was extinct in Britain in 1900, recovered to 15 in 1958 but was down to one breeding pair at Minsmere in 1971. The birds were badly affected by pesticides and other pollutants throughout Europe. Over the last 30 years the increase has been spectacular – almost 20% a year – possibly over 200 by now. The stronghold is still in the bigger reed beds in East Anglia but now many pairs use smaller ones and arable crops. Breeding is now common round the Wash and in the

Fens with regular widespread records in other areas including Wales. They are now breeding for the first time in Scotland and are back in Ireland after about 150 years. Many are migrants, wintering in Africa, and seem to wander widely and may settle to breed far from their natal areas. *Prospects good – could become a familiar bird in farmland as well as wetland areas.*
Underhill-Day, J. 1998 *British Birds*: 91, 210-218.

HEN HARRIER

 *Circus cyaneus*

> *Distribution* Britain 498 (+33.5%) Ireland 123 (-35.9%)
> *Numbers breeding*: Britain 630 Ireland 180
> *European status*: 9,500 (9% in Britain and Ireland =4)
> *British population trend*: good – if persecution stops!
> *How likely are you to record it?* 30 squares (0.7%) *Ranked* 129= [77=]

These birds were widespread 150 years ago but lost their breeding habitat and were destroyed by gamekeepers. Although they have recovered they are now being shamefully restricted by illegal persecution, mainly on grouse moors. Some populations are doing well (e.g. Isle of Man) but, recently, the thriving Orkney population has diminished. In Ireland (about 100 pairs in 1998), and elsewhere, declines may be because of maturing forestry plantations – new plantings are often ideal nesting sites. The Langholm report (see Persecution and Protection) showed that they may increase where heather moorland is overgrazed by sheep. It is then replaced by grass and the Meadow Pipits (and other small birds) increase, providing excellent feeding for the Hen Harriers in spring. These conditions are very bad for the Red Grouse. Average of 16.4 breeding females in Wales (1988-94) with nest failures down to predation (foxes, crows, keepers). Breeding numbers would very possibly be doubled if the illegal persecution were to cease. *A gradual increase possible, if illegal persecution is reduced, with breeding at coastal marshes of SE England.*
Meek, E.R. *et al.* 1998 *Scottish Birds*: 19, 290-299.
Potts, G.R. 1998 *Ibis*: 140, 76-88.
Ridpath, S.M. & Thirgood, S.J. 1997 *Birds of prey and red grouse*. London: Stationery Office.

PALLID HARRIER *Circus macrourus*

A male on Orkney paired with a Hen Harrier in 1995 but the nest, with 5 eggs, failed – an amazing and unexpected record.

MONTAGU'S HARRIER

Circus pygargus

> *Distribution* Britain 32 (-36.0%) Ireland 0
> *Numbers breeding*: Britain 13 RBBP Ireland 0
> *European status*: 8,000 (0% in Britain and Ireland)
> *British population trend*: gradual increase possible
> *How likely are you to record it?* 2 squares (0.0%) *Ranked* 183=

Always rare over the last 100 years, about 30 pairs nested in the 1950s on moorland and heathland (up to 10 pairs in Wales: none after the mid-1960s). Numbers then dwindled and none was recorded nesting in 1974. The last Irish nest was in 1971. A gradual return started almost immediately and in the last ten years RBBP had logged 6 – 12 breeding females and between 1 and 11 non-breeding birds. Most are in East Anglia and around the New Forest. Some of these migrant birds have nested in arable crops – where success is high when careful protection is provided. Global warming may help them. *Slow increase possible but not assured.*

GOSHAWK (Northern Goshawk)

Accipiter gentilis

> Distribution Britain 236 (+574%) Ireland 1
> Numbers breeding: Britain 347 RBBP Ireland 0
> European status: 75,000 (0% in Britain and Ireland)
> British population trend: further increase
> How likely are you to record it? 20 squares (0.4%) Ranked 136=

The Goshawk's history is blanketed by a fog of dis-information. It supposedly became extinct 100 years ago and sporadic breeding started about 70 years ago – fuelled by deliberate releases and escaped falconer's birds. Certainly by 25 years ago the real population was probably quite a bit bigger than the official 'the total certainly exceeds ten pairs'. Now it is breeding very widely through Wales, possibly 150 or more pairs by now, and the situation in Scotland is not much different. In England the distribution is more patchy as the extensive woodland they need is a rarer resource. Maturing conifer forests will inevitably be colonised. They were lost from Ireland with the clearance of the forest 200 years ago but a pair nested recently in the North. Illegal persecution is the main problem. *Further expansion expected to go ahead.*

SPARROWHAWK (Eurasian Sparrowhawk)

Accipiter nisus

> Distribution Britain 2,178 (+19.6%) Ireland 663 (-17.0%)
> Numbers breeding: Britain 32,000 – Ireland 11,000
> European status: 155,000 (28% in Britain and Ireland =1)
> British population trend: recovered – now stable? (+294% CBC){+162%}
> How likely are you to record it? 600 squares (13.7%) Ranked 58 [57=]

This scourge of the small woodland birds now breeds throughout Britain and Ireland, except the Western Isles, Shetland and other areas without trees. This was the situation to 1950 despite the best attempts of gamekeepers to destroy them. The only of success was that the populations in some areas increased during the two world wars, when many keepers were away on other duties. In the late 1950s it became clear that they had crashed and were missing from large areas, particularly of South-east England. The culprits were the organo-chlorine pesticides like Dieldrin, Aldrin and DDT, and their metabolites, which built up in the food chain and poisoned the top predators. This caused many deaths and disrupted the female's calcium metabolism leading to many failed breeding attempts. The chemicals were banned and the birds made good most of their losses (not yet complete in the South-east). They also took to the new conifer forests, where they have to put up with Goshawk predation, and started breeding in towns. *Full recovery seems likely.* Newton, I., Dale. L. & Rothery, P. 1997 *Bird Study*: **44**, 129-135.

BUZZARD (Common Buzzard)

Buteo buteo

Distribution Britain 1,544 (+8.4%) Ireland 93 (+272%)
Numbers breeding: Britain 14,500 Ireland 150 +
European status: 410,000 (4% in Britain and Ireland =8)
British population trend: expanding quite rapidly (+29% CBC){+224%}
How likely are you to record it? 1009 squares (22.5%) *Ranked* 47[84=]

The mewing of the Buzzard is now a traditional sound of upland Britain. Two centuries ago it was everywhere and in the northern part of Ireland but, considered as vermin, it was extinguished from all lowland England (save the New Forest) and Ireland and rendered rare elsewhere, even Wales, by 1910. Then it gradually began to increase and returned (1933) to Ireland – up to 10 pairs by 1954 – but then the dearth of rabbits (myxomatosis) caused another extinction by 1964. Buzzards were soon back and now may number 200 pairs. Apart from the lack of rabbits, organo-chlorine pesticides and illegal persecution have been problems – and the latter persists. However a steady, but slow, expansion east in England seems to be accelerating and this bird looks set to take its rightful place, alongside the Red Kite, as a widespread large bird of prey over most of the country. *Prospects excellent.*

GOLDEN EAGLE

Aquila chrysaetos

Distribution Britain 408 (+5.4%) Ireland 0
Numbers breeding: Britain 422 Ireland 0
European status: 5,400 (8% in Britain and Ireland =4)
British population trend: stable
How likely are you to record it? 9 squares (0.2%) *Ranked* 151=

This huge predator with its vast home range has never been common but was very severely persecuted by gamekeepers and shepherds – and trophy collectors. They were gone from Wales by about 1750. In Ireland breeding stopped in 1912 with only an unsuccessful pair in Antrim since (1950s). A pair (sometimes 2) have tried to nest in England since 1960 after an absence of about 180 years. However this is a bird of the Highlands of Scotland where there are 422 pairs (1992) – almost static since 1982 (424) although the area occupied has altered. The species gradually recovered as legal protection was observed but illegal persecution and egg collecting still happen. Some forty years ago the birds were also affected by toxic chemicals, mainly from sheep dips. Carrion, rabbits and hares are probably in good supply. They may spread to other parts of the southern Scotland, expand in northern England and return to Ireland where a reintroduction scheme is due to start in Donegal. *Set fair?*

OSPREY

Pandion haliaetus

> *Distribution* Britain 168 (+572%) Ireland 2 (0%)
> *Numbers breeding*: Britain 99 RBBP Ireland 0
> *European status*: 5,000 (2% in Britain and Ireland =7)
> *British population trend*: further expansion assured
> *How likely are you to record it?* 9 squares (0.2%) *Ranked* 151=

This was the flagship species for nature conservation following its progressive extermination from Scotland, mainly by collectors. The last nesting attempts were in 1908 and 1916. Birds returned and nested in 1954 and gradual establishment and expansion ensued. They are popular with visitors and often a status symbol for the local lairds. One is alleged to have stocked the pond in front of his house with trout for the local breeding pair! Many sites used are now traditional with nests growing bigger and bigger each year. The total of pairs with nests reached 111 in 1997 with a steady increase (doubling every ten years). About 90% of the pairs breed and the 75% that are successful raise, on average, two young. In recent years chicks from Scottish nests have been released at Rutland Water and, in 1999 birds from this scheme returned to summer. Within a year or two breeding in England will start again after a lapse of about 200 years – but will Rutland or a wild pair elsewhere be first? They may even breed again in Ireland after an absence of 250 years. Shooting in Africa still accounts for some British birds. *Up to 400 pairs, all over the place, by 2025?*

KESTREL (Common Kestrel)

Falco tinnunculus

> *Distribution* Britain 2,481 (-4.1%) Ireland 804 (-15.0%)
> *Numbers breeding*: Britain 50,000 Ireland 10,000
> *European status*: 280,000 (21% in Britain and Ireland =1)
> *British population trend*: declining -24% CBC (-18% BBS){-17%}
> *How likely are you to record it?* 1157 squares (25.8%) *Ranked* 41 [40]

The hovering Kestrel, head to wind, searching for small mammals beside a busy road seems to indicate the integration of its old lifestyle with modern conditions. This may not be a true picture. The birds are found just about everywhere except Shetland, Harris and Lewis. They had been much reduced by keepers 100 years ago but, even before they received protection, many keepers had realised they were not a real threat as they mainly eat small mammals. The restored population suffered declines, blamed on seed-dressings, and fluctuates in many areas from year to year with the cyclic changes in vole numbers. Indeed earlier catastrophic vole plagues in Scotland (1880s) could have been due to the lack of control from this and other avian predators. The most recent Atlas records losses in Ireland and South-west England and Scotland and CBC and BBS record declines but brood sizes are increasing and nest losses declining. The population declines are of serious concern and may indicate that the small mammals are not doing well. Lack and loss of nest sites may exclude Kestrels from some areas and boxes erected on poles are readily used in treeless areas. *There is serious cause for concern.*

MERLIN

Falco columbarius

> *Distribution* Britain 693 (+15.9%) Ireland 158 (-35.5%)
> *Numbers breeding*: Britain 1,300 Ireland 120
> *European status*: 13,000 (6% in Britain and Ireland =5)
> *British population trend*: doing quite well
> *How likely are you to record it?* 46 squares (1.0%) *Ranked* 116= [76=]

This dashing little falcon of moorland was an easy target for gamekeepers with pole-traps. The few lowland breeding sites in England had been lost almost 150 years ago. The populations elsewhere in Britain were reduced except for the very north of Scotland and, possibly, Wales. They were back on Exmoor and Dartmoor by 1920 but the overall population

did not recover like the other raptors and the birds were hit hard by chemical pollution. The new Atlas confirmed a recent retreat in Ireland and the loss of most birds from the south-west of England and some from Wales. However breeding birds are now using the safer abandoned nests of crows in forestry rather than nesting on the ground and there are definite increases in some areas – for instance Northumbria. The Irish population may still be decreasing but, overall, the British population increased between 1983/84 and 1993/94 particularly, more than doubling in northern England. Possible country totals are Scotland 800, England 400 and Wales 100 pairs. *Has the Merlin's time come? Let's hope so!*
Rebecca, G.W. & Bainbridge, I.P. 1998 *Bird Study*: **45**, 172-181.
Williams, I.T. & Parr, S.J. 1995 *Welsh Birds*: 1, 14-20.

HOBBY (Eurasian Hobby)

Falco subbuteo

> *Distribution* Britain 625 (+141%) Ireland 3 (0%)
> *Numbers breeding*: Britain 624 RBBP Ireland 0
> *European status*: 21,000 (3% in Britain and Ireland =10)
> *British population trend*: continued rapid increase
> *How likely are you to record it?* 58 squares (1.3%) *Ranked* 111

The migrant Hobby is not a danger to gamebirds but nonetheless was persecuted in the past – either by mistake for Sparrowhawk or simply as a 'hooky bill'. It was thus depleted at the turn of the century but most counties south of a line from the Humber to the Severn had breeding birds. The birds retreated south and by the time of the first Breeding Atlas they were thought to be birds of open heathland and were mostly breeding south of the Thames and about 50 km north of it. There were thought to be about 100 pairs. Then it was realised that many were breeding in farmland and they were mapped into Wales and as far north as Shropshire, South Yorkshire and even Tyne and Wear in the second Atlas! The RBBP records indicate a doubling in about 12 years and 1997 was a record year with up to 624 pairs recorded. In Ireland birds are present most years in Wicklow and colonisation is eagerly awaited.
Further expansion and more expected.

PEREGRINE (Peregrine Falcon)

Falco peregrinus

> *Distribution* Britain 1,048 (+104%) Ireland 290 (+166%)
> *Numbers breeding*: Britain 1,185 Ireland 365
> *European status*: 5,800 (27% in Britain and Ireland =2)
> *British population trend*: more than ever and in new areas
> *How likely are you to record it?* 60 squares (1.3%) *Ranked* 110 [71=]

Who would have thought, 40 years ago, that there would be record numbers of Peregrines breeding in Britain. These magnificent birds were virtually farmed in some areas, to provide birds for falconry for several centuries. However they began to be persecuted, particularly on grouse moors, in the 19th century but seemed to have a relatively stable population. Emergency legislation turned everyone against them during the Second World War – they might intercept homing pigeons used for emergency messages. Over 600 were killed and many nests destroyed with a result that the English (worst affected) population was reduced by over 50%. After the war the birds gradually recovered, probably to the pre-war level, in ten years. Then the organo-chlorine chemicals were introduced into agricultural use. Peregrine numbers started falling and breeding failures were reported. At the same time racing pigeon fanciers complained about the losses of their birds and the first census was carried out by the BTO in 1961-62. Numbers were down to half the pre-war population! The chemicals were withdrawn and eventually banned and the sample surveys indicated that Peregrines were in their worst state in 1963 with 44% of the pre-war population and only 16% breeding successfully. Further censuses, both in Britain and Ireland, indicated the numbers were made up by 1980 in most areas. Numbers and breeding performance in areas where the major part of the diet was seabirds were not good but elsewhere many traditional territories now held two or three pairs. The estimate for 1991 was over 1,500 – a third up on pre-war figures. Apart from a few on church spires, well over 100 years ago, lowland areas did not have breeding Peregrines but this is now changing with breeding increasingly being reported on pylons, cooling towers, warehouses and even blocks of high rise flats. The pigeons in Trafalgar Square need to watch out as at least one pair nested in London in 1998! *Is 2,000 pairs the limit?*

Crick, H.Q.P. & Ratcliffe, D.A. 1995 *Bird Study*: **42**, 1-19.

Moore, N.P. *et al.* 1997 *Bird Study*: **44**, 176-181.

Norriss, D.W. 1995 *Bird Study*: **42**, 20-30.

Ridpath, S.M. & Thirgood, S.J. 1997 *Birds of prey and red grouse*. London: Stationery Office.

RED GROUSE (Willow Ptarmigan)

Lagopus lagopus

> *Distribution* Britain 945 (-12.7%) Ireland 141 (-66.4%)
> *Numbers breeding*: Britain 250,000 Ireland 3,000
> *European status*: 1,200,000 (23% in Britain and Ireland =2)
> *British population trend:* probably poor but up 28% on BBS
> *How likely are you to record it?* 242 squares (5.4%) *Ranked 85 [75=]*

The Red Grouse used to be considered the only truly British species but it is now relegated to sub-specific status of the circumpolar Willow Grouse. These birds were, and still are in some areas, very important gamebirds with very careful management of their heather moorlands by burning and predator control. Populations and shooting bags were at the highest at the end of the 19th century – and have gone downhill since. Lack of management during the First World War, and the Second, over grazing by sheep, conversion of heather to grass or forestry have all contributed. Declines in bags, on well managed moors, has been about 50% over the last 60 years. Grouse populations exhibit marked cycles over 4 to 8 years which are caused by nematode parasites. The collapse of numbers can be stopped by catching and dosing a good percentage of the breeding birds with vermicides in spring. In Ireland densities of birds are less than in Britain and the birds have declined and contracted in range very seriously over the last 80 years – even though they do not exhibit the same cycles as the British birds. *Further declines seem inevitable where serious money is not available for management.*
Ridpath, S.M. & Thirgood, S.J. 1997 *Birds of prey and red grouse.* London: Stationery Office.

PTARMIGAN (Rock Ptarmigan)

Lagopus mutus

> *Distribution* Britain 173 (-11.3%) Ireland 0
> *Numbers breeding*: Britain 10,000 Ireland 0
> *European status*: 530,000 (2% in Britain and Ireland =5)
> *British population trend:* apparently stable
> *How likely are you to record it?* 5 squares (0.1%) *Ranked 162=*

These amazing birds live in the very special Arctic-alpine habitat only found in Scotland in the Highlands – although actually down to about 200 m above sea level in the far north-west. These birds bred, more than

100 years ago, on mountains in southern Scotland and on Hoy (Orkney). On the Western Isles they were eventually lost from their last foothold on Lewis in 1938. They do not require the heather that the Red Grouse need and are generally found above them on the hill where they graze such plants as crowberry and blaeberry. In the winter their plumage changes from the brown, with which they are superbly camouflaged in the snow-free summer, to white so that they disappear in the snow. Losses have been caused in many areas through degradation of the habitat through overgrazing by sheep. Around the tourist areas on Cairngorm the edible rubbish which has been discarded has encouraged Crows and foxes to the extent that predation has increased greatly. This is one of the species at risk from global warming as its habitat may, in the long term, be squeezed from the Scottish scene. *Short term prospects reasonable.*

BLACK GROUSE

Tetrao tetrix

> *Distribution* Britain 432 (-28.4%) Ireland 0
> *Numbers breeding*: Britain 15,000 Ireland 0
> *European status*: 700,000 (2% in Britain and Ireland =5)
> *British population trend*: retreating and declining badly
> *How likely are you to record it?* 23 squares (0.5%) *Ranked* 133=

Few other birds have shown such a consistent loss in Britain. Birdwatchers in London could find leks within a few dozens of miles of home a century ago! Some may have come from introductions but the last birds were shot in Kent and Sussex in the late 19th century, in Surrey, Berkshire and Norfolk in 1905, in Hampshire, Wiltshire and Dorset by 1928 and Lincolnshire a few years later. For a short while there seemed to be an increase put down to their ability to exploit newly planted conifers but, alas, losses then continued with Devon birds gone by the time of the second Atlas. In Wales the decrease of males from 1986 to 1995 was 38% – to 164. Early 1999 has probably seen the demise the species from the Peak District. In Northern Britain the game bag records seem to indicate that 95% have been lost over the last century and there have certainly been severe problems over much of Scotland. The British population could be almost in free fall at the moment (estimated at 15,000). However careful management, including predator control, habitat manipulation and the removal of deer fences (see Capercaillie) has reversed the trend in some areas. *These birds are in serious trouble.*

Baines, D. & Hudson, P.J. 1995 *Bird Study*: **42**, 122-131.
Hancock, M. *et al.* 1999. *Bird Study*: 46, 1- 15.
Williams, I.T. 1996 *Welsh Birds*: 1, 22-28.

CAPERCAILLIE (Western Capercaillie)

🐾 🐾 ☹ ☹   *Tetrao urogallus*

Distribution Britain 66 (-63.7%) Ireland 0
Numbers breeding: Britain 2,200 Ireland 0
European status: 250,000 (1% in Britain and Ireland)
British population trend: deep and desperate decline
How likely are you to record it? 2 squares (0.0%) *Ranked* 183=

These are *seriously* big birds, the males weigh 4 kg, and they are, for the second time, in serious trouble – they were extinct from 1785 in Scotland and rather earlier from Ireland. Birds were successfully introduced, from Sweden to Scotland, in 1837 and 1838 and for several more years. They reached their maximum about 90 years ago and remained fairly static for 40 years or so. There was then a further increase linked to the planting of new conifers and, by the time of the first Atlas, they occupied much of the high forested ground between the Great Glen and the Lowlands. Since then the population has crashed and the estimate of 2,200 may be very optimistic. Degradation of the habitat through overgrazing by sheep and deer and increased predation, mainly by the expanding fox population, are important factors. However attention has recently focused on deer fences. These are up to almost three metres high and cause deaths to flying grouse which would normally expect to be able to blast their way through the outer branches of trees. Grants are now available to remove old fences and to make useful fences obvious to the birds. The high deer populations, of

course, cause the problem. *Let's hope the help is in time.*
UKBAP **Scottish Natural Heritage** Institute for Terrestrial Ecology & RSPB.

BOB-WHITE QUAIL (Northern Bobwhite)

Colinus virginianus

These little North American gamebirds have been introduced for shooting many times in Britain (and at least once in Ireland) with success lasting no more than 20 years. Two populations (Scilly and Suffolk coast) were breeding at the time of the first Breeding Atlas but there were only records from one 10-km square during the second.

CHUKAR *Alectoris chukar*

These close relatives of the Red-legged Partridge, and hybrids between the two, were released from 1970 onwards to 'improve' the stock. They did not and both hybrid and pure pairs were much less productive – their release has been illegal since 1992.

RED-LEGGED PARTRIDGE

Alectoris rufa

Distribution Britain 1,214 (+32.1%) Ireland 12
Numbers breeding: Britain 170,000 territories Ireland 0
European status: 3,300,000 (5% in Britain and Ireland =3)
British population trend: fairly good but -30% CBC (+19% BBS){-6%}
How likely are you to record it? 846 squares (18.8%) *Ranked* 54 [101=]

This south-western European species was successfully introduced, for sport, to Suffolk in 1770, after a 100 years of unsuccessful attempts. The eggs, reared under chickens came from France – hence the alternative name of French Partridge. The birds gradually expanded to reach something like their current extent by 1930 but the exact situation is always complicated by new releases – even tried (unsuccessfully) in Ireland and in many places in Scotland (some resulting in breeding). Some 30 years ago Chukars (see above) and hybrids were released muddying the situation but now this is banned. Recent declines may be related to new patterns of release but the species may do well with hotter summers through global warming. *Could do very well.*

GREY PARTRIDGE

Perdix perdix

> *Distribution* Britain 1,629 (-18.7%) Ireland 35 (-86.3%)
> *Numbers breeding*: Britain 145,000 — Ireland 50
> *European status*: 2,000,000 (7% in Britain and Ireland =3)
> *British population trend*: steep decline (-78% CBC){-82%}
> *How likely are you to record it?* 531 squares (11.8%) *Ranked* 60

This native gamebird is, for most sportsmen, the best. It prospered during the last century, but arsenical seed-dressings took their toll from 200 years ago! Introductions and rearing complicate the records but weather related problems, harsh winters and wet summers, knocked the population back. Losses in Ireland have been catastrophic and from being widespread it may become extinct soon. Indeed in 1999 there were thought to be only about 20 breeding pairs restricted to two sites and they may, without drastic intervention, be lost in the next few years. It has also been lost from much of Wales, the south-west peninsula and parts of Scotland. Detailed research by the GCT (Game Conservancy Trust)shows this is due to the lack of the insect food for the young chicks

due to the intensification of agriculture. Paradoxically the birds are doing well in areas like Norfolk, where they are intensively shot but well looked after. Perhaps they should be taken off the game list on estates where they are *not* carefully nurtured! *Prospects not good except in core areas.* UKBAP MAFF Game Conservancy Trust.

QUAIL (Common Quail)

Coturnix coturnix

> *Distribution* Britain 804 (+98.5%) Ireland 34 (+3.0%)
> *Numbers breeding*: Britain 300 Ireland 20
> *European status*: 720,000 (0% in Britain and Ireland)
> *British population trend*: sporadic – good year during Atlas
> *How likely are you to record it?* 56 squares (1.2%) *Ranked* 113 [101=]

Quails are tiny, migratory gamebirds whose breeding population fluctuates wildly in our area. Huge numbers are shot on the Continent and this may be the reason for a general decline at the end of the 19th century. Now, in good years which are not very frequent, Quail can be found anywhere suitable for them to breed. Recent Quail years have been 1964, 1970, 1884 and 1989 – when there may have been 2,600 calling males in Britain and 90 in Ireland. In a normal year there may be 100-300 pairs in Britain and fewer than 20 in Ireland. *Mixed prospects!*

SILVER PHEASANT *Lophura nycthemera*

Introductions have never become self-sustaining but this species is often seen as an escape.
Ogilvie, M.A. & RBBP 1999 *British Birds*: **92**, 176-182 & 472-476.

REEVE'S PHEASANT *Syrmaticus pictus*

Transitory attempts at introduction – the latest on Speyside in about 1970 – have been made. During the first Atlas they were recorded near Woburn (Bedfordshire), in Breckland and on Speyside. Now it is likely that the occasional records of single birds in Britain all refer to recent escapes.
Ogilvie, M.A. & RBBP 1999 *British Birds*: **92**, 176-182 & 472-476.

PHEASANT (Common Pheasant)

Phasianus colchicus

> *Distribution* Britain 2,269 (+0.9%) Ireland 844 (-0.7%)
> *Numbers breeding*: Britain 1,550,000 + Ireland 570,000 females
> *European status*: 4,100,000 (51% in Britain and Ireland =1)
> *British population trend*: past and future assured (+83% CBC){+65%}
> *How likely are you to record it?* 2704 squares (60.2%) *Ranked* 17 [12]

This common introduced gamebird has shaped the management of many areas within the British countryside. It has become widespread and numerous over most of Britain and Ireland – save Shetland and much of the Hebrides and is extensively reared. About 200,000 are shot annually in Ireland and more than 20,000,000 in Britain! Many are released birds, no more a part of the natural avifauna than battery chickens. On some estates over 90,000 birds are released onto ordinary farmland at rates approaching 25 per hectare, with terrible problems of disease and predation. However often the habitat is managed for the Pheasants, with great benefits for other species, and high populations of 'wild' birds are bred and topped up with releases of just one bird per hectare. The 'Ring-necked' birds and other varieties, like 'Blue-backs', are all the same species. *An important rural industry.*

GOLDEN PHEASANT

Chrysolophus pictus

> *Distribution* Britain 47 (+80.8%) Ireland 0
> *Numbers breeding*: Britain 750 Ireland 0
> *European status*: 750 (100% in Britain and Ireland =1)
> *British population trend*: slowly dwindling
> *How likely are you to record it?* 5 squares (0.1%) *Ranked* 162=

This splendid but secretive bird was introduced from China over 100 years ago to Dumfries and Norfolk and later Sussex, Hampshire, Mull, Anglesey and elsewhere. Useless as a gamebird, it is established in several places but has suffered declines in recent years – even in its Breckland stronghold. *Cause for concern: if one should be concerned about an exotic species.*

Ogilvie, M.A. & RBBP 1999 *British Birds*: **92**, 176-182 & 472-476.

LADY AMHERST'S PHEASANT

Chrysolophus amherstiae

> *Distribution* Britain 9 (-30.8%) Ireland 0
> *Numbers breeding*: Britain 50 RBBP individuals Ireland 0
> *European status*: 75 (100% in Britain and Ireland =1)
> *British population trend*: possibly on the way out?

This is another very secretive woodland pheasant in several places but particularly successfully on the Bedfordshire/Buckinghamshire border. The population here seems to be in decline and may be less than 100 birds. There may be a handful of birds elsewhere – e.g. Sussex, Hampshire, Norfolk and Cheshire. *More concern of the same sort!*
Ogilvie, M.A. & RBBP 1999 *British Birds*: **92**, 176-182 & 472-476.

WATER RAIL

Rallus aquaticus

> *Distribution* Britain 420 (-33.6%) Ireland 176 (-36.5%)
> *Numbers breeding*: Britain 700 Ireland 1,300
> *European status*: 160,000 (1% in Britain and Ireland)
> *British population trend*: assured but may be declining
> *How likely are you to record it?* 5 squares (0.1%) *Ranked* 162= [101]

These are very difficult birds to record and even worse to prove breeding – they have amazing black balls of fluff for young chicks. The best bet is to hear them 'sharming' (calling) in the evening or at night – like a stuck pig! In the 19th century they were breeding everywhere save the Northern and Western Isles and the eggs were regularly collected for the table. But, in Britain but not so much in Ireland, the birds thinned out and disappeared from many areas with increasing drainage. There were further losses between the Atlases *but* crepuscular birds, like this, may have been under-recorded. This is one of the species that needs further study by enthusiasts, using the newly formulated census techniques now available, to find out what is actually happening. *Many breed in protected areas and at least these have reasonable prospects.*
Jenkins, R. 1999 *BTO News*: **221**, 7.

SPOTTED CRAKE

Porzana porzana

> *Distribution* Britain 26 (-33.3%) Ireland 1 (0%)
> *Numbers breeding*: Britain 14 RBBP Ireland 0
> *European status*: 55,000 (0% in Britain and Ireland)
> *British population trend*: just hanging on!

These birds *may* formerly have been widespread and quite common in Britain and even have bred regularly in Ireland 150 or 200 years ago. Now they are widely scattered and infrequent save at major wetland sites in Inverness and Cambridgeshire. Seldom seen, they are usually recorded by call at night. Only three or four birds were recorded in several years in the early 1980s but a recent peak (1993) was of 31 recorded males and there were 14 in 1997. *Just a small toehold.*
Adam, R.G. & Booth, C.J. 1999 *Scottish Birds*: 20, 14-17.

BAILLON'S CRAKE *Porzana pusilla*

East Anglian breeding proved a few times in the middle of the 19th century (1858-1889) but never even suspected since. However there was an obliging bird calling in Kent in summer 1999.

CORNCRAKE (Corn Crake)

Crex crex

> *Distribution* Britain 161 (-75.6%) Ireland 246 (-70.3%)
> *Numbers breeding*: Britain 641 RBBP Ireland 158
> *European status*: 90,000 (2% in Britain and Ireland =10)
> *British population trend*: huge decline turned round recently
> *How likely are you to record it?* 6 squares (0.1%) *Ranked* 160=

The loss of this bird is well documented and clearly caused by changes to agriculture. It breeds in rank, tangled vegetation – particularly hay meadows. Towards the end of the 19th century it had started to decline as machine mowing of hay started. The hay was cut early so that the birds were unable to raise young and, later, silage was even worse. The declines accelerated and during the fieldwork for the first Atlas addition retreats were noted (especially Ireland) and sporadic records – mainly of birds calling their scientific name *crex* – in mainland Britain. The second Atlas has only 18 records in England and Wales over the four years. They were confined to the Western Islands of Scotland, Orkney and parts

of western Ireland. Corncrake recovery schemes were formulated, with Government funding, in both Scotland and Ireland. The farmers agreed emergency changes to the management of their hay crops and the declines have been halted, and reversed, just in time! In 1997 there were 637 in the core Scottish area, 4 (!) elsewhere in Britain and 3 in N. Ireland and in 1998 the Republic had 2(!) outside the core area with 149-153. *There is still a long way to go but they are coming back from the brink!*
Casey, C. 1998 *Irish Birds*: **6**, 159-176.
Green, R.E. 1995 *Bird Study*: **42**, 66-75.
UKBAP **Scottish Office** The Scottish Office & RSPB.

MOORHEN (Common Moorhen)

Gallinula chloropus

> *Distribution* Britain 2,032 (-9.2%) Ireland 714 (-21.0%)
> *Numbers breeding*: Britain 240,000 Ireland 75,000 territories
> *European status*: 1,000,000 (32% in Britain and Ireland =1)
> *British population trend*: slight decline (-10% CBC, +27% WBS){-26%}
> *How likely are you to record it?* 1041 squares (23.2%) *Ranked* 46 [56]

These birds can be very obvious and tame but in some areas they may be timid and crake-like. They can use almost any sort of pool, pond, stream or river and are all over the place except the north and west of Scotland and some parts of western Ireland. They are vulnerable to cold winter weather and seem to have lost 50% of their breeding population after the exceptional cold of 1962/63. There is good evidence that they are vulnerable to predation by feral mink. There is a decline registered by the CBC but an increase by the WBS – this may be related to the loss of farm ponds and other water as the decline was most marked on farmland CBC (-18%). *The future seems assured.*

COOT (Common Coot)

Fulica atra

> *Distribution* Britain 1,603 (-5.1%) Ireland 354 (-37.2%)
> *Numbers breeding*: Britain 46,000 Ireland 8,600 adults
> *European status*: 1,200,000 (2% in Britain and Ireland)
> *British population trend*: relatively stable – CBC down, WBS up{+17%}
> *How likely are you to record it?* 411 squares (9.2%) *Ranked* 66 [68]

Coots need bodies of shallow, open and vegetated water on which to breed and, where they can find it, they will be present. They are absent

Coot and Moorhen.

from much of North-west Scotland and the islands – but recently increased on Orkney. There were declines 50 or 60 years ago and there is evidence that they were lost from some squares between the two breeding atlases – especially in Ireland. However there are very good numbers of birds at many breeding sites and although the CBC index fell by 22% the WBS increased by 274% – both were possibly unrepresentative but the difference between the two is that WBS *have* to be wet! In many areas Coots have become tame birds of urban parks largely feeding on bread! *Here to stay.*

CRANE (Common Crane)

 Grus grus

Extinct by the 17ᵗʰ century, from many areas (including Ireland) and finally East Anglia, surprisingly Cranes settled there again in 1981. There were three pairs in 1997 and one youngster was raised – the first since 1988.

GREAT BUSTARD *Otis tarda*

Extinct in Britain about 160 years ago, an attempt at re-introduction to Salisbury Plain failed some years ago. One remaining male from this attempt was still at Whipsnade Zoo in 1998. Further re-introductions to Wiltshire are planned.

OYSTERCATCHER (Eurasian Oystercatcher)

Haematopus ostralegus

> *Distribution* Britain 1,702 (+11.4%) Ireland 263 (+0.8%)
> *Numbers breeding*: Britain 38,000 Ireland 3,500
> *European status*: 235,000 (18% in Britain and Ireland =2)
> *British population trend*: long term increases (-16% BBS)
> *How likely are you to record it?* 510 squares (11.4%) *Ranked 63 [75=]*

The black and white Oystercatcher is a big, bold, noisy and conspicuous bird. It used to be exclusively a coastal breeder (not East coast of Ireland) but seems to have started to breed inland on the north coast of Grampian about 160 years ago. River nesting, on shingle banks, and nesting on islands and the shores of lakes spread in this area, along the north side of the Solway and around Lough Erne by the end of the century. In Ireland the East Coast was colonised about 75 years ago but they are still not nesting along about 100 km of the South Coast. Inland breeding, save for the few round Lough Erne, is still very unusual. However breeding inland has steadily spread to encompass virtually all Scotland, Northern England, many East Anglian river valleys, parts of the Midlands and started in Wales about 25 years ago. Breeding birds have been lost from Devon and Cornwall. Changes in behaviour have been suggested for the expansion but ringing shows that the birds are now very long-lived and, if this is new, it may be the reason. Roof nesting is recorded – possible since the adults bring food to their chicks (unlike most waders). Winter food availability may eventually limit the population growth. *Further increases confidently expected.*

BLACK-WINGED STILT

Himantopus himantopus

Breeding, some successful, recorded in Nottinghamshire (1945: 3 pairs), Gwent (early 1950s), Cambridgeshire (1983), Norfolk (1987) and Cheshire (1993). Sammy, a lone male, has haunted Titchwell (Norfolk) for several years. *They could turn up, probably in East Anglia, any time!*

AVOCET (Pied Avocet)

Recurvirostra avosetta

> *Distribution* Britain 28 (+250%) Ireland 0
> *Numbers breeding*: Britain 654 RBBP Ireland 0
> *European status*: 28,000 (2% in Britain and Ireland)
> *British population trend*: spreading and increasing
> *How likely are you to record it?* 8 squares (0.2%) *Ranked* 155=

This species was chosen as the RSPB logo when they started to return to breed: courtesy of Hitler! The defence of the East Anglian coast involved the flooding of the coastal marshes and provided the birds with ideal breeding habitat. The records are not complete but breeding took place in Norfolk and Essex and in 1947 there were small colonies both at Havergate and Minsmere (Suffolk). The birds had bred in the early 1800s from the Humber round to Sussex with one or two later records (over 100 years ago) in Suffolk. A record of two successful breeding pairs in Wexford in 1938 was amazing and not repeated! After the initial post-war breeding the numbers only built up at Havergate almost reaching 100 in 1958. However, by 1963 when the Minsmere birds first started to breed regularly, there were fewer than 60 pairs between the two sites. Spasmodic nesting elsewhere was recorded until the early 1970s when the population reached about 150 (about 75% at Havergate and 25% at Minsmere). Now they breed from Yorkshire to Sussex and the number of pairs reported to RBBP seem to be gradually increasing reaching 654 by 1997. Breeding is now taking place inland on the Ouse Washes (first recorded in 1996 and up to 20 pairs by 1999) and the birds are summering at Rutland Water. The return to Britain has happened after many years of increase in the number of Avocets breeding in Denmark, Holland and other countries round the North Sea – for instance the Danish population trebled over the 50 years 1920 – 1970 and the Dutch population increased even faster. *Provision and management of the habitat is the answer.*

STONE CURLEW (Stone-curlew)

Burhinus oedicnemus

> *Distribution* Britain 54 (-41.9%) Ireland 0
> *Numbers breeding*: Britain 203 RBBP Ireland 0
> *European status*: 37,000 (0% in Britain and Ireland)
> *British population trend*: decline turned round in last few years
> *How likely are you to record it?* 9 squares (0.2%) *Ranked* 151=

The Stone Curlew is really a bird of dry steppe and is right on the edge of its range in Britain. Historically there are records from Dorset through Gloucestershire, Worcestershire and North to eastern Yorkshire but the edges of this range had begun to lose their birds before the end of the 19th century. The loss of heathland to arable was blamed and birds were down in numbers in many areas. The birds fluctuated as crops and cropping patterns changed but the main trend was downwards. The farming slump 70 years ago caused temporary increase but the Yorkshire birds went in 1938. Marginal land was planted with conifer plantations and the depredations of myxomatosis badly affected the birds – a lack of rabbits meant that bare areas of ground became overgrown. During the first Atlas they were not found in Lincolnshire but were still breeding along the South Downs, at Dungeness, along the Suffolk coast and in the two main areas – North Norfolk through Breckland and down the Chilterns and the downland from Berkshire, through Hampshire and Wiltshire into Dorset. There may have been as many as 500 pairs. The second Atlas saw further contraction, although there was a female on eggs in Lincolnshire, the population reached a low of 150-160 pairs or less. Very careful and detailed RSPB research has discovered what the birds need and how to deliver it so the population is now increasing and a minimum of 233 pairs bred in 1999. The BDAP target for 2000, 200 pairs, was reached by 1998. *Prospects excellent whilst the money to look after them is forthcoming.*

Bealey, C.E. *et al.* 1999 *Bird Study*: **46**, 145-156.

UKBAP **MAFF** RSPB.

LITTLE RINGED PLOVER (Little Plover)

Charadrius dubius

> *Distribution* Britain 421 (+46.2%) Ireland 1 (0%)
> *Numbers breeding*: Britain 950 Ireland 0
> *European status*: 75,000 (1% in Britain and Ireland)
> *British population trend*: still increasing and expanding
> *How likely are you to record it?* 23 squares (0.5%) *Ranked* 133=

The historic record of breeding in Britain by this small inland plover is easy – it didn't! In 1938 a pair of these birds, hitherto rare migrants in Britain, nested at Tring Reservoirs and then, in 1944, there were two and another pair in Middlesex. They have not looked back. Estimates of the population explosion are 100 pairs by 1958, 200 by 1964 and about 400 by 1973. The birds were nesting east of a line from Selsey Bill to the Mersey and North to the Scottish Border. The birds mainly used gravel pits, reservoir and lake shores, sewage works and the gravel beds associated with larger rivers. Now they will even nest in crops or setaside in sandy and gravelly areas like Breckland. Birds have consolidated and extended their range west – but although there are over 50 pairs in Wales, they are still missing from the north-west. Breeding has only occurred about dozen times in Scotland but they should soon become regular and may colonise Ireland quite soon. One cloud on the horizon is the reported displacement of this species from traditional breeding sites by its larger cousin, the Ringed Plover, as they expand inland. *Great potential for doubling or trebling the population.*

RINGED PLOVER

Charadrius hiaticula

> *Distribution* Britain 1,025 (+12.2%) Ireland 244 (-21.5%)
> *Numbers breeding*: Britain 8,500 Ireland 1,250
> *European status*: 95,000 (10% in Britain and Ireland =4)
> *British population trend*: further slow expansion likely
> *How likely are you to record it?* 45 squares (1.0%) *Ranked* 118=

Until quite recently this species was considered to be a bird of coastal sand and shingle. However it has always nested on shingle spits and stony shores of rivers and lochs in Scotland, Northumberland and inland at the major Irish loughs. A thriving population breeding in the rabbit warrens of the Breckland of Norfolk and Suffolk was severely depleted by 1900. In many areas where tourists spent their holidays the disturbance

banished the birds from their favourite sites and the overall population probably continued to fall to a minimum about 30 years ago. Then conservation considerations started to help the coastal birds and increasing numbers started to breed inland. These were in the Brecklands, along the Thames and Trent and in many other areas including extensive areas in Lowland Scotland. However inland breeding has still not come to Wales and may be declining in Ireland – but some are now breeding amid the desolation of stripped peat! There are still some coastal areas of southern Ireland, Wales and the South-west peninsula from which they are absent. *More inland breeding is very likely.*
Cooney, T. 1998 *Irish Birds*: 6, 283-284.

KENTISH PLOVER

 Charadrius alexandrinus

This bird, named after an English county, is extinct in Britain. Many, perhaps hundreds, of pairs bred in Kent and Sussex 150 years ago but regular breeding, following much persecution, ceased in about 1931. Few sporadic records since then with the last in Lincolnshire 1979. The regular nesting on the Channel Islands ceased in 1975. *Outlook continuing grim.*

DOTTEREL (Eurasian Dotterel)

 Charadrius morinellus

Distribution Britain 99 (+115%) Ireland 0
Numbers breeding: Britain 900 Ireland 0
European status: 25,000 (4% in Britain and Ireland =4)
British population trend: more still being found
How likely are you to record it? 3 squares (6.7%) *Ranked* 171=

These are birds of the High Tops and not easy to survey. About 150 years ago there may have been over 50 pairs in England, mainly the Lake District, but these dwindled and may have died out at the turn of the century only to return a few years later in small numbers. The Highland population also went through a very bad period: birds were killed for gourmet food and for their feathers (fly-tying). Breeding resumed in South-west Scotland in 1967 and in North Wales in 1969. Between 60 and 100 pairs were estimated for the first Atlas. Improved survey techniques and more birds, by the time of the second Atlas, provided an estimate of 950 pairs. Birds (not proved to breed) were recorded in North Wales and there was a breeding record in Ireland (Mayo 1975). Recent RBBP records indicate irregular breeding in England (and very irregular

in Wales); however the core area seems to have a steady and thriving population. *Global warming may squeeze this species out in the future: the only cloud on the horizon.*
Galbraith, H. *et al.* 1993 *Bird Study*: 40, 161-169.
Strowger, J. 1998 *Bird Study*: 45, 85-91.

GOLDEN PLOVER (European Golden Plover)

Pluvialis apricaria

> *Distribution* Britain 784 (-7.7%) Ireland 57 (-13.6%)
> *Numbers breeding*: Britain 22,600 Ireland 400
> *European status*: 550,000 (4% in Britain and Ireland =5)
> *British population trend*: serious declines set to continue
> *How likely are you to record it?* 188 squares (4.2%) *Ranked* 90 [77=]

In the 19th century these moorland breeding birds bred on uplands through most of Scotland, Northern England, Wales (especially the North), Ireland and Exmoor. The birds then retreated from the edges and populations fluctuated in the core areas. Breeding on Dartmoor was a gain – and eventually the only area in the south-west with breeding birds. The Irish birds did badly and were restricted to Connemara and Donegal by 1960: the Atlas results show further declines. The Welsh and Dartmoor populations are seriously depleted. Worries about increased disturbance from hill walkers do not seem to be borne out everywhere. Overgrazing and the deterioration of the moorland habitats could be a cause but the birds also winter on arable farmland. This means that the intensification of agriculture, in the wintering areas, may be a very real problem. Certainly a study in the Flow country, where there is hardly any disturbance and no sign of degradation of the breeding habitat, shows serious recent declines. These may be caused by deteriorating conditions in lowland areas where they winter. *There is grave concern about these birds throughout Europe.*
Brown, A.F. 1993 *Bird Study*: 40, 196-202.
Hancock, M. & Avery, M. 1998 *Scottish Birds*: 19, 195-205.

LAPWING (Northern Lapwing)

Vanellus vanellus

> *Distribution* Britain 2,340 (-9.0%) Ireland 491 (-28.8%)
> *Numbers breeding*: Britain 126,300 Ireland 21,500
> *European status*: 1,300,000 (19% in Britain and Ireland = 1=)
> *British population trend*: steep decline – halved in 11 years!{-52%}
> *How likely are you to record it?* 1108 squares (24.7%) *Ranked* 42 [60]

The Lapwing, Peewit or Green Plover is, or has been, a familiar bird of open farmland, marshes and moors throughout Britain and Ireland. During the 19th century the populations declined due to drainage and widespread and organised egg collecting for food. A Lapwing protection bill (1926) was passed and caused a temporary increase but the decline continued. Ireland, especially the south-west, was badly affected, especially by losses in cold winters, like 1962/63. There have also been considerable declines on the Shannon Callows, one of the core areas in Ireland. Recent BTO and RSPB research documented a 47% loss of breeding birds from England and Wales in just 11 years! Wales (down about 73%) and South-west England (down about 64%) have fared worst but much of the rest of England has lost over half their breeding birds. The stronghold is in the North where two thirds of the remaining birds are breeding and the losses are (only) a bit over 40%. Loss of breeding sites and low productivity are implicated in the rapid decline, due to the intensification of agriculture and particularly the change from spring to autumn sowing. *Outlook seems to be fairly grim but should be OK on reserves.*

TEMMINCK'S STINT

Calidris temminckii

A handful of breeding attempts in the Scottish highlands from 1934 culminated in proof during 1971 and now there are generally two or three pairs each year: slightly more between 1974 and 1981 with up to 5 pairs and 4 sites.

PURPLE SANDPIPER

Calidris maritima

Breeding of these colonists on high ground in Scotland first proved in 1978. Birds have been present in one or two (widely separated) sites each year since – maximum possibly four pairs 1981, 1989 and 1991.

DUNLIN

Calidris alpina

> *Distribution* Britain 569 (+20.6%) Ireland 69 (+7.8%)
> *Numbers breeding*: Britain 9,500 Ireland 175
> *European status*: 350,000 (3% in Britain and Ireland =4)
> *British population trend*: recent declines reported
> *How likely are you to record it?* 58 squares (1.3%) *Ranked* 111= [93=]

The most common wader on our winter estuaries but the breeding population is less than 10,000 pairs. There were more in the 19th century with lowland breeding birds in some areas of Ireland, along the coast of Cheshire, Lancashire, the Solway and Fife – even on heathland in Lincolnshire. These birds had mostly gone by 1950 but there are very significant and dense populations in the Western Isles on the machair. Other upland populations are sparser and some have suffered from afforestation. The few Irish birds, breeding evidence in only 22 squares in second Atlas compared with 39 in the first, are divided between the small amount of local machair and upland sites. Breeding birds in Devon, Cornwall, Wales, northern England and South-west Scotland are in steep (terminal?) decline. The recent discovery of feral mink on North Uist together with egg eating by hedgehogs is bad news. However, there are recent reports is that there may be more breeding in remote areas than previously estimated. *Some cause for concern.*

Hancock, M. & Avery, M. 1998 *Scottish Birds*: **19**, 195-205.

Rae, R. & Watson, A. 1998 *Scottish Birds*: **19**, 185-195

RUFF

 Philomachus pugnax

Extinct as a regular breeder from eastern counties in England 1850; there were sporadic records to 1922. Return started in 1963 on the Ouse Washes and up to 10 leks reported – most in traditional areas but some in Anglesey, North-west England and Scotland (only proved once) but fewer recently. *Coming soon?*

SNIPE (Common Snipe)

Gallinago gallinago

> *Distribution* Britain 1,806 (-19.1%) Ireland 641 (-28.4%)
> *Numbers breeding*: Britain 55,000 — Ireland 10,000 +
> *European status*: 920,000 (8% in Britain and Ireland =5)
> *British population trend*: contracting and declining (-90% CBC){-74%}
> *How likely are you to record it?* 245 squares (5.5%) *Ranked* 84 [46]

Snipe breed in upland areas and on lowland marsh lands. Here the population was reduced by drainage 100 years ago but they were able to recover some ground as agricultural land went out of production during the slump. About 50 years ago lowland population again began a downward plunge, still in progress, with birds missing from large areas of lowland England and Wales by the time of the first Atlas – much worse for the second with losses also in Ireland (probably stable in the

Shannon Callows) and lowland Scotland. In some areas upland populations seem also to be affected. The modern high stocking rates on grazing land are not helpful as there will be an increased risk of nests being trampled – nest records seem to show this is happening. Withdrawal from the shooting list might seem a good idea but most of those shot here are wintering birds from overseas. If the resources the birds need for breeding are restricting the population, this is unlikely to make much of a real difference. *Lowland outlook very poor with the restoration of wet areas to low intensity grazing the only hope.*

WOODCOCK (Eurasian Woodcock)

 Scolopax rusticola

Distribution Britain 1,204 (-28.9%) Ireland 179 (-64.1%)
Numbers breeding: Britain 15,000 Ireland 3,000
European status: 600,000 (3% in Britain and Ireland = 8=)
British population trend: seriously declining (-55% CBC){-70%}
How likely are you to record it? 8 squares (0.2%) *Ranked* 155= [101=]

The squeaky, croaking whistles of roding (displaying male) Woodcock were not at all familiar until about 1820 when the species started to breed extensively in Britain and Ireland. Within 10 or 20 years it had spread to breed in almost all counties – possibly as a result of the protection of woodland where Pheasants were nesting and increased planting of trees. The population seemed stable and well established at the time of the first Atlas and during 1975 survey in Ireland. But widespread losses were recorded in the second Atlas (1988-1991): crepuscular species were probably less well recorded. However there is very good evidence for a real, widespread and substantial decrease over the last 15 or 20 years. Maturing forests and dry springs have been blamed but Ireland and the West are wet and most of the area of loss appears not to have altered much. However many of the pastures, where they feed, have been improved and powerful chemicals are used to dose sheep and cattle against worms and other parasites. *Outlook again poor – are we going back 100 years?*

BLACK-TAILED GODWIT

Limosa limosa

> *Distribution* Britain 59 (+37.2%) Ireland 9 (+50.0%)
> *Numbers breeding*: Britain 47 Ireland 2
> *European status*: 145,000 (0% in Britain and Ireland)
> *British population trend*: set to increase
> *How likely are you to record it?* 3 squares (0.1%) *Ranked* 171=

These large and handsome waders had ceased to breed regularly in eastern counties from Yorkshire to Suffolk, by 1855. Sporadic nesting restarted in 1934 and regularly at the Ouse Washes in 1952 (Nene Washes 25 years later). Others have nested in various marshes from Northumberland to Kent, on the Somerset Levels and in the North-west possibly from Anglesey to the Solway. Currently the annual total is 40 to 50 pairs. These are probably the nominate race *limosa* but a few Icelandic birds (*islandica*) have bred in Northern Scotland (regularly Shetland), in the last 40 years. The first breeding record in Ireland was 1975 and scattered pairs (*islandica*) continue to nest regularly. *Prospects seem pretty good from Iceland and the Continent.*

WHIMBREL

Numenius phaeopus

> *Distribution* Britain 83 (+40.7%) Ireland 0
> *Numbers breeding*: Britain 530 Ireland 0
> *European status*: 230,000 (0% in Britain and Ireland)
> *British population trend:* increasing well mostly in Shetland
> *How likely are you to record it?* 42 squares (0.9%) *Ranked* 121 [70]

This smaller cousin of the Curlew, familiar to many birdwatchers through its 'seven whistlers' migration call, only breeds in the far North. It had declined to 30 pairs and was confined to Shetland by 1900. A gradual increase started about 70 years ago and the population had doubled by about 1960. The increase possibly accelerated and a survey in the 1980s estimated there were about 450 pairs on Shetland (70 on Fetlar alone) and 20-25 elsewhere in Scotland – regular on Orkney and sporadic on St Kilda, Lewis and the northern Mainland. A complete Shetland census found 479 pairs in 1994. In the future predation by Great Skuas may be a problem. *1,000 pairs by 2010?*
Dore, C.P., Ellis, P.M. & Stuart, E.M. 1996 *Scottish Birds*: 18, 193-196.

CURLEW (Eurasian Curlew)

Numenius arquata

> *Distribution* Britain 1,893 (-2.8%) Ireland 671 (-19.6%)
> *Numbers breeding*: Britain 35,500 Ireland 12,000
> *European status*: 135,000 (36% in Britain and Ireland =2)
> *British population trend*: declining (-26% CBC){-28%}
> *How likely are you to record it?* 927 squares (20.6%) *Ranked* 52 [44]

Until 150 years ago these big waders bred only on upland moorland but then they started to colonise lowland areas. By 1900 they were nesting on Exmoor, the Lancashire Mosses, the New Forest Heaths and some downland. Further expansion took in river valleys and even agricultural land in areas like the Midlands. In the Scottish heartland the birds started to nest in lower areas over 70 years ago and now regularly nest in arable crops. In Ireland the birds were not common but they were widely distributed and apparently stable in the Shannon Callows. In recent years there have been widespread reports of declines possibly due to drainage and the effects of agricultural intensification and predation. However this is not likely to be responsible for a very obvious loss from the south-eastern half of Ireland, the south-west peninsula and midlands of England and Pembrokeshire between the two atlases. *Outlook not at all good.*

REDSHANK (Common Redshank)

Tringa totanus

> *Distribution* Britain 1,473 (-11.8%) Ireland 213 (-15.1%)
> *Numbers breeding*: Britain 32,100 – Ireland 4,700
> *European status*: 350,000 (11% in Britain and Ireland =4)
> *British population trend*: losing out in many areas (-72% CBC){-60%}
> *How likely are you to record it?* 127 squares (2.8%) *Ranked* 98 [88=]

This nervous and noisy wader, the sentinel of the marshes, requires wet marshland (fresh or salt) to breed – almost colonial in favoured areas. About 1850 they declined but came back from 1870 for 70 years. The birds are vulnerable to cold winter weather and drainage. In Ireland breeding birds are found only in the Shannon Callows (severe recent declines), around Lough Erne and Lough Neagh with a handful in the machair of the North-west and some in Wexford and Wicklow. In England most are in the North-west, coastal marshes from Hampshire to the Humber and inland in Hampshire and parts of the Midlands. In the West the Somerset Levels and Severn Basin retain their birds but they are declining and very sparse in Wales. In Scotland they are absent from much of the Highlands but very dense on the machairs of the West. Further losses seem inevitable and even the hot spots on the machair may be at risk from mink and hedgehogs and on salt marshes from rising sea-levels and increased grazing pressures. *Outlook pretty poor in many traditional areas.*

GREENSHANK (Common Greenshank)

Tringa nebularia

> *Distribution* Britain 243 (-4.0%) Ireland 1 (0%)
> *Numbers breeding*: Britain 1,350 Ireland 0
> *European status*: 70,000 (2% in Britain and Ireland =4)
> *British population trend:* seems stable and secure
> *How likely are you to record it?* 32 squares (0.7%) *Ranked* 125= [101=]

This very elegant and rare wader was only found to be breeding in Scotland about 180 years ago and immediately became a prized bird for egg collectors. Eventually it was realised that the birds were mostly breeding in the higher marshes of the bigger rivers, west of the Great Glen. The upper reaches of the Spey were colonised about 100 years ago – possibly following forest clearance – and birds bred, and breed, in the Hebrides. One pair was known in Ireland in the early 1970s and they may breed irregularly now. *The future seems secure.*

GREEN SANDPIPER *Tringa ochropus*

Only twice proved to breed – Westmorland 1917 and Speyside 1959 – but a pair reported at the same site in Highland during 1995 and 1996 and strongly suspected to have bred.

WOOD SANDPIPER

Tringa glareola

Sporadic breeding in England about 1850 associated with the now extinct Dutch population. First confirmed Highland Scotland 1959 and now up to 15 pairs may breed at 9 sites (1997 RBBP). *Increasing again for last five years.*

COMMON SANDPIPER

Actitis hypoleucos

> *Distribution* Britain 1,424 (+1.3%) Ireland 313 (-30.8%)
> *Numbers breeding*: Britain 15,800 Ireland 2,300
> *European status*: 570,000 (3% in Britain and Ireland =5)
> *British population trend*: still declining? (-16% WBS){-14%}
> *How likely are you to record it?* 154 squares (3.4%) *Ranked* 95 [71=]

Over a century ago this familiar wader, on migration, was a regular breeder in most lowland counties but it had started to withdraw to the uplands as early as 1850 in Dorset. However even 100 years ago they still bred in parts of Sussex and many areas along lowland rivers. In Ireland there has been a steady contraction to the north and west and most are now in Donegal and Connemara – with a cluster in Kerry and a few may still breed in the Wicklow mountains. In Scotland, northern England and Wales, where there are upland streams, the birds are still common but there is some evidence of decline over the last 50 or 60 years. The declines are probably continuing in Ireland and lowland parts of Grampian but the birds have returned to breed in several places in Devon and Cornwall. Apart from these records they remain absent from south and east of a line from the Tees to Bristol. Acidification of the rivers and streams has led to losses of food resources. *There may be further losses.*

SPOTTED SANDPIPER *Actitis macularia*

A pair of this vagrant North American species, closely related to the Common Sandpiper, attempted to breed on Skye in 1975. The first breeding record for Europe (one probably paired with a Common Sandpiper) produced young in Yorkshire in 1991!

TURNSTONE (Ruddy Turnstone)

Arenaria interpres

These birds often summer around our coasts and are sometimes seen to make scrapes and even copulate – most often in the northern islands. In 1976 they may have bred in Sutherland.

RED-NECKED PHALAROPE

🐎 🐎 🍀 🍀 ☹

Phalaropus lobatus

> *Distribution* Britain 9 (-50.0%) Ireland 1 (-75.0%)
> *Numbers breeding*: Britain 38 RBBP Ireland 1
> *European status*: 80,000 (0% in Britain and Ireland)
> *British population trend*: very precarious

These delicate waders are found on and around shallow pools in the far north and west. Early records imply breeding in several places on mainland Scotland but were much persecuted by collectors. By 1900 the Western Isles, Orkney and Shetland were the only regular breeding sites but they were also discovered in Mayo. The subsequent peak of 50 pairs was rather short lived and the Irish population dwindled to a handful – up to 25 for 1967-71 – and, despite a few records away from Mayo, now seldom register more than one or two pairs. In Scotland regular breeding on Orkney stopped 25 years ago, it is only very sporadic on the mainland and very few breed now on the Western Isles. Their stronghold is Shetland, and particularly Fetlar, where 37 males raised at least 54 young in 1997 through careful habitat management by the RSPB. *Very worrying effectively having only two breeding sites.*

ARCTIC SKUA

☹

Stercorarius parasiticus

> *Distribution* Britain 113 (+14.1%) Ireland 0
> *Numbers breeding*: Britain 3,200 Ireland 0
> *European status*: 20,000 (17% in Britain and Ireland =3)
> *British population trend*: possibly continued losses in prospect
> *How likely are you to record it?* 13 squares (0.3%) *Ranked* 144

Early writers reported these birds from very much the same areas as they now breed – the Western Isles, Orkney and Shetland and some on the northern tip of mainland Scotland. However they were severely persecuted by gamekeepers and collectors, and at a low ebb 100 years ago, particularly in Caithness and Orkney. With better protection they slowly increased, suffering some setbacks in Shetland where they lost out in competition with the expanding and bigger Great Skua. For a few years, in the 1960s, a bird (or two) summered in Ireland. An estimate of 1,090 breeding pairs in 1969-70 was followed by 3,350 in 1985-87! However almost 2,000 of these were on Shetland where breeding failures and declines were then recorded at many sites, possibly due to sand-eel failures and Great Skua predation. For instance at Foula the high of 280 pairs in 1976 was down by about 57% in 1996 and 1997. *Further losses seem very likely.*

LONG-TAILED SKUA

Stercorarius longicaudus

Individuals are sometimes seen at Arctic Skua colonies and they may have bred at a mainland site in 1980. There were records of adults in Grampian (1974) and Speyside (1975) in June.

GREAT SKUA

Catharacta skua

> *Distribution* Britain 97 (+38.6%) Ireland 0
> *Numbers breeding*: Britain 8,500 Ireland 0
> *European status*: 13,700 (58% in Britain and Ireland =1)
> *British population trend:* increasing and expanding
> *How likely are you to record it?* 11 squares (0.2%) *Ranked* 148=

These birds probably first colonised Scotland in Shetland (where they are known as Bonxies) in about 1750. At first they were welcome as they kept predators away from the sheep and lambs but they were the target of collectors through the next 150 years and there were still few breeding by 1900. Since then they have been protected and the numbers may have doubled every decade until 1970! They now breed on Shetland (ca. 90%), Orkney, the Western Isles and the North and North-west coast of the mainland. The expansion southwards looks set to continue gradually but the breeding success was so bad on Shetland that the population there declined for a while about 10 years ago. The birds are now doing better and breeding numbers are increasing as well as, rather significantly, the non-breeding birds in the club areas. These birds may be serious predators of other seabirds (see Marine and Coastal) and even hazard the continued existence of colonies of small petrels at sites like St Kilda – where 145 pairs in 1994 had increased to 271 by 1997! There may be problems ahead for them as new European regulations about discards from trawlers restrict their 'free lunches'. However these are very resourceful birds. *Very much on the up and up.*

Phillips, R.A., Thompson, D.R. & Hamer, K.C. 1999 *J. Appl. Ecol.*: **36**, 218-232.
Phillips, R.A. *et al.* 1999 *Bird Study*: **46**, 174-183.

MEDITERRANEAN GULL

 Larus melanocephalus

Distribution Britain 7 (+600%) Ireland 0
Numbers breeding: Britain 53 RBBP Ireland 0
European status: 250,000 (0% in Britain and Ireland)
British population trend: new and increasing quite rapidly
How likely are you to record it? 1 square (0.0%) *Ranked* 190=

After rapid expansion from the Black Sea this gull first bred in Hampshire in 1968 and proper pairs and hybrids with Black-headed Gulls were then recorded from more and more sites – reaching 10 in 1990 and up to 53 pairs at 24 sites in 1997. Birds have now been recorded in Black-headed Gull colonies in Scotland and the first successful breeding in Ireland was in Wexford in 1996 , although eggs were laid in Antrim in 1995. *Here to stay.*
Allen, D. & Tickner, M. 1996 *Irish Birds*: 5, 435-436.

LITTLE GULL

Larus minutus

Little Gulls are sporadic breeders outside their normal range – in all four pairs laid eggs in 1975, 1978 (two pairs) and 1987 in England, and one or two pairs probably bred in Scotland in 1991.

BLACK-HEADED GULL

 Larus ridibundus

Distribution Britain 671 (-18.8%) Ireland 145 (-48.4%)
Numbers breeding: Britain 167,000 Ireland 53,800
European status: 2,100,000 (11% in Britain and Ireland = 3=)
British population trend: fairly stable (-27% BBS)
How likely are you to record it? 952 squares (21.2%) *Ranked* 50 [52]

The Black-headed Gull is a very familiar bird all over the country but breeds in rather few places in large colonies. A hundred years ago the effect of egg collecting for food had reduced the colonies in areas with good communications but the birds were doing well in remoter upland sites. At this time the colonies in Southern Britain, coastal and inland, started to increase – about 35,000 pairs in England and Wales rose to about 100,000 in 1973. Eggs are still collected for food but many are from tidal areas and would have been lost to spring high tides. In Scotland

there has been recent range expansion and the population there and in Ireland has probably been increasing from 100 years until about 30 years ago. Since then the situation is rather confused as colonies may shift, exchange birds with neighbours or remain uncounted for years. As human egging becomes less important, mammalian predation by foxes, hedgehogs and mink is becoming worse. Several colonies have over 10,000 pairs. *The future seems assured.*

Whilde, A., Cotton, D.C.F. & Sheppard, J.R. 1993 *Irish Birds*: 5, 67-72.

COMMON GULL (Mew Gull)

Larus canus

Distribution Britain 577 (-12.7%) Ireland 87 (-39.2%)
Numbers breeding: Britain 68,000 Ireland 3,600
European status: 475,000 (15% in Britain and Ireland =3)
British population trend: gradual losses (+16% BBS)
How likely are you to record it? 264 squares (5.9%) *Ranked* 78 [66]

There is considerable confusion with early records as the Common Gull was not the commonest gull but, by 1900, it is certain that it bred in the far north-west of Ireland and over most of Scotland bar the south-east corner. Over the next 60 years they spread with colonies in much of the north-west half of Ireland, in new areas in Scotland and established at Dungeness, Kent, 1919, with desultory breeding in northern England. Anglesey was colonised in about 1960 but the breeding birds were foxed to extinction 10 years later. More recently many colonies have declined or disappeared, although they are now established inland in northern England. Afforestation and predation by mink and foxes have been implicated but there are still some large, thriving and dispersed inland colonies. *The losses seem likely to continue.*

Whilde, A., Cotton, D.C.F. & Sheppard, J.R. 1993 *Irish Birds*: 5, 67-72.

LESSER BLACK-BACKED GULL

 Larus fuscus

> *Distribution* Britain 434 (+0.5%) Ireland 81 (-31.4%)
> *Numbers breeding*: Britain 83,000 Ireland 5,200
> *European status*: 220,000 (40% in Britain and Ireland =1)
> *British population trend*: increased but now declining (+39% BBS)
> *How likely are you to record it?* 940 squares (20.9%) *Ranked* 51 [54]

In common with many large birds, perceived to compete with man for sporting birds, these birds were persecuted during the 19th century and exterminated from some places near centres of population. They increased as game keeping declined and the birds were probably helped a lot by discards from trawlers. Overall the birds spread and increased until perhaps 20 years ago, although this was not true in Orkney and Shetland – possibly due to Great Skuas. Many colonies, some larger than 10,000 breeding pairs, were culled (often to protect populations of other birds) and the surviving birds redistributed themselves. By 1994 it was estimated that 3,200 pairs bred on buildings, an increase of 17% *per annum* since 1976. Populations, overall, now seem to be in decline and productivity is poor in some areas. Britain and Ireland hold 40% of the European population and they are doing badly in North Norway. *Serious cause for concern.*

Raven, S.J. & Coulson, J.C. 1997. *Bird Study*: 44, 13-34.
Whilde, A., Cotton, D.C.F. & Sheppard, J.R. 1993 *Irish Birds*: 5, 67-72.

HERRING GULL

Larus argentatus

> *Distribution* Britain 729 (-1.1%) Ireland 163 (-27.6%)
> *Numbers breeding*: Britain 160,000 Ireland 44,700
> *European status*: 790,000 (26% in Britain and Ireland =1)
> *British population trend*: down 40% in 20 years (+33% BBS)
> *How likely are you to record it?* 1055 squares (23.5%) *Ranked* 46 [37]

Herring Gulls were well established round our coasts long before they started to appear in the introduction to *Desert Island Discs*! Populations started to increase in the early 1900s and reached a maximum about 30 years ago. Discards from fisheries, better general bird protection, rubbish management that allowed the gulls many free lunches were all invoked as reasons for the increases. Between 1976 and 1994 breeding on buildings increased by 10% *per annum* to almost 17,000 pairs. However the overall coastal population is thought to have declined from 283,900 to 146,700 pairs round Britain and 59,700 to 44,200 pairs round Ireland in the years 1969 to 1987. Culling for all sorts of reasons, restriction of feeding opportunities provided by man and the effects of botulism, mink and fox predation may all have had, and are probably still having, their effects on this widespread species. *Not, surely, in serious trouble?*
Raven, S.J. & Coulson, J.C. 1997. *Bird Study*: **44**, 13-34.
Whilde, A., Cotton, D.C.F. & Sheppard, J.R. 1993 *Irish Birds*: **5**, 67-72.

YELLOW-LEGGED GULL

Larus cachinnans

This newly separated species first appeared breeding as part of a hybrid pair, with Lesser Black-backed Gull, in 1992. In 1995 a pair bred successfully in Dorset and there were two pairs there in 1997. *Definitely a coming species!*

GREAT BLACK-BACKED GULL

Larus marinus

> *Distribution* Britain 486 (-4.4%) Ireland 137 (-28.3%)
> *Numbers breeding*: Britain 19,000 Ireland 4,500
> *European status*: 105,000 (22% in Britain and Ireland =2)
> *British population trend*: probably stable (+7% BBS)
> *How likely are you to record it?* 212 squares (4.7%) *Ranked* 86 [57=]

This big and impressive species was much persecuted both as a predator and also for trophies and was in a parlous state by about 1875 when its fortunes started to improve. Better legal protection was the spur and they started to spread around the rocky coasts – with a handful breeding inland. Some are colonial, and they mainly feed on fish and crustaceans but there are also solitary pairs that specialise on predating other seabirds. In Ireland the increase seems to have slowed or stopped in the North 30 years ago but it continued elsewhere. In Scotland the overall increase has probably been reversed by the sharp declines in Shetland and Orkney probably attributable to harassment or competition from the increasing Great Skuas – about half the national total nested on these islands in 1969-70. *Prospects good – except where there are Bonxies!*
Whilde, A., Cotton, D.C.F. & Sheppard, J.R. 1993 *Irish Birds*: 5, 67-72.

KITTIWAKE (Black-legged Kittiwake)

Rissa tridactyla

Distribution Britain 252 (+4.2%) Ireland 62 (-20.5%)
Numbers breeding: Britain 490,000 Ireland 50,200
European status: 2,350,000 (23% in Britain and Ireland =3)
British population trend: possibly declining now after increases
How likely are you to record it? 3 squares (0.1%) *Ranked* 171= [93]

These delicate cliff-nesting gulls were shot in very large numbers during the 19th century for sport (!) but particularly for the millinery trade. This largely ceased following the 1869 Seabirds Preservation Act and the depleted colonies, mainly the few in England, Wales and easily accessible parts of Scotland began to recover. For 50 years – between 1920 and 1969 – there was an annual increase of about 3.5%, but it then slowed. Since 1980 or so there have been many reports of declines, although a few increases have been reported, with Shetland colonies (of major

importance) halved between 1981 and 1997. Here problems with sand-eel stocks and severe predation by Great Skuas may be responsible. However the return to South-east England in the last three decades is encouraging and there were 3,760 pairs counted in 1985-87: a huge increase on the 80 pairs found in 1969. Even after the current changes these birds are much more common than they were 50 years ago. *Overall continued decline likely as Great Skuas increase in what were their stronghold islands.*

Heubeck, M., Mellor, R.M. & Harvey, P.V. 1997 *Seabird*: **19**, 12-21.
Heubeck, M. *et al.* 1999 *Bird Study*: **46**, 48-61.
McGrath, D. & Walsh, P.M. 1996 *Irish Birds*: 5, 375-380.

GULL-BILLED TERN *Sterna nilotica*

A pair was proved to breed at Abberton Reservoir, Essex, in 1950 and had been present the previous year too.

LESSER CRESTED TERN
Sterna bengalensis

A long-time favourite, and star of TV programmes, Elsie has paired and produced several young with a male Sandwich Tern on the Farne Islands, Northumbria from 1984. One of these hybrids has also returned there to breed with a Sandwich Tern.

SANDWICH TERN

Sterna sandvicensis

> *Distribution* Britain 43 (-24.6%) Ireland 37 (+8.8%)
> *Numbers breeding*: Britain 14,000 Ireland 4,400
> *European status*: 125,000 (15% in Britain and Ireland =3)
> *British population trend*: recent losses – outlook uncertain
> *How likely are you to record it?* 15 squares (0.3%) *Ranked* 140= [88=]

A hundred years ago this species was in a parlous state as its eggs were sold for food. There were only a few dozen breeding in colonies on Walney, Moray and Nairn and Orkney but over 100 in Co. Mayo and on the Farnes. In Ireland many new areas were colonised with 3,500 pairs in 1984, 4,400 pairs about five years later but only 2,941 in 1995. The Scottish situation was similar with two well established colonies on the Sands of Forvie (near Aberdeen) and Inchmickery (Firth of Forth) holding over 2,000 pairs; there were a few hundred at other sites, particularly Orkney. Recently the two main colonies have gone and less than 1,000 pairs now breed. In Wales up to 1,000 pairs now breed on Anglesey. In

England about 20 colonies are now regular (biggest in Northumbria and Norfolk) with over 10,000 pairs in all. These birds are sometimes very mobile and shift breeding sites but killing, by children in West Africa, may be important in limiting numbers. *Are recent losses a warning for the future?*
Hannon, C., Berrow, S.D. & Newton, S.F. 1997 *Irish Birds*: 6, 1-22.

ROSEATE TERN

 *Sterna dougallii*

Distribution Britain 19 (-17.4%) Ireland 11 (-50.0%)
Numbers breeding: Britain 55 RBBP Ireland 649 RBBP
European status: 1,610 (31% in Britain and Ireland =2)
British population trend: very good – at Rockabill!
How likely are you to record it? 2 squares (0.0%) *Ranked* 183=

This is a species with a terrible history of human exploitation – first on the breeding grounds and latterly in winter by West African children. By 1900 there were only a handful breeding – most off Anglesey and none in Ireland. Here the population started to recover and was regularly several hundred pairs, in scattered colonies, and a huge colony on a sandy island in Wexford Harbour was discovered – 2,000 pairs in 1962. Erosion destroyed this site and Ireland held 268 pairs of the 525 total European population (excluding the Azores) in 1984! Now careful conservation at Rockabill, off Dublin, and Lady's Island Lake (Wexford) has consolidated the population at 611 and 116 nests in 1999. In Scotland over 500 pairs sometimes bred (e.g. 450 in the Firth of Forth 40 years ago) but now the total is generally about 10 pairs. In Wales the Anglesey colonies were up to 250 about 30 years ago but now staggers on at 10 or so despite recent intensive EU funded management of the nesting area. The English colonies did number several hundred pairs on the islands off Northumberland but there are now only a few dozen at scattered sites. The trapping in West Africa is a problem being worked on but the breeding birds seem to move in response to their prey availability. *The Irish experience will enable us to prepare five star breeding sites for the birds – if they can find a good food resource!*
Hannon, C., Berrow, S.D. & Newton, S.F. 1997 *Irish Birds*: 6, 1-22.
UKBAP **English Nature & DoE(NI) – Environment and Heritage Service**
RSPB

COMMON TERN

Sterna   hirundo

> *Distribution* Britain 426 (-7.0%) Ireland 98 (-43.4%)
> *Numbers breeding*: Britain 12,300 Ireland 3,100
> *European status*: 210,000 (8% in Britain and Ireland =4)
> *British population trend*: probably quite good
> *How likely are you to record it?* 102 squares (2.3%) *Ranked* [77=]

The Common Tern is the commonest tern over the southern half of Britain but is out-numbered by the Arctic in the northern half of the country. It is another species that was badly affected by predation for millinery (mainly the wings) and eggs for food. By 1900 there were rather few colonies round the English coasts, in Anglesey in Wales but more round Scotland and Ireland (inland also in the northern and western parts). The Irish birds were thought to far outnumber Arctic Terns on the East coast earlier this century but this was reversed in South and West. By 1970 the Common/Arctic ratio was 2.8:1 but the species had reached parity on the coast by 1984 with Common still outnumbering Arctic by six to one inland. Numbers were 2,848 in 1984 and 3,053 in 1995. In Scotland the Isle of May had over 5,000 pairs in the mid-1940s but it was extinguished by the expanding colony of large gulls. After culling the gulls, there are now several hundred pairs and about 4,000 in Scotland. Mink predation exacts a terrible toll in areas like Argyll, Harris and Lewis. In Wales the birds only breed regularly in Anglesey and Flint. The English population has gradually expanded inland and on artificial sites where rafts have been made for them. *Prospects probably good in areas where they are looked after.*
Hannon, C., Berrow, S.D. & Newton, S.F. 1997 *Irish Birds*: 6, 1-22.

ARCTIC TERN

Sterna   paradisaea

> *Distribution* Britain 303 (-15.7%) Ireland 72 (-19.1%)
> *Numbers breeding*: Britain 44,000 Ireland 2,500
> *European status*: 520,000 (9% in Britain and Ireland =2)
> *British population trend*: bad, largely because of mink
> *How likely are you to record it?* 12 squares (0.3%) *Ranked* 145= [93=]

This species is the northerly member of the 'comic' group whose general distribution in Britain and Ireland has changed little in the last hundred years. Apart from substantial colonies on the Northumberland Islands and Ynys Mon (Anglesey) there are generally only a few dozen pairs breeding south of the Lowlands of Scotland within bigger colonies of Common Terns. Early in the 19th century there had been a good number on the Scillies and only 30 years ago 200 pairs at the entrance to Morecambe Bay. The Irish situation is described under Common Tern (above) but the numbers are encouraging with 1,041 (1969/70 coastal colonies only), 2,460 (1984 all) and 3,092 (1995 all). The Scottish colonies illustrate the fickle nature of the species with huge shifts from year to year. For example there were 750 on Foula in 1973; 1,800 in 1975 and 6,000 in 1976 and 400 on nearby Papa Stour in 1969-79, 3,000 ten years later (1978-80) and then 10,000 in 1981. However all is not well. There have been massive breeding failures over the last 15 years in Shetland and the population has seriously declined, probably due to the inability of the birds to catch sufficient (any?) sand-eels. Mink and global warming pose further problems. *The long-term prospects for Arctic Terns do not look at all good.*

Hannon, C., Berrow, S.D. & Newton, S.F. 1997 *Irish Birds*: 6, 1-22.

LITTLE TERN

Sterna albifrons

Distribution Britain 110 (-18.5%) Ireland 36 (-30.8%)
Numbers breeding: Britain 2,400 Ireland 390
European status: 21,500 (13% in Britain and Ireland =2)
British population trend: better because of protection
How likely are you to record it? 4 squares (0.1%) *Ranked 167=*

The Little Tern, with its very well camouflaged eggs and beach nesting habit, can easily be trampled out of existence at colonies where human feet intrude. As if this was not hazard enough some colonies become seaside burger bars for predators like foxes and Kestrels. A hundred years ago the Little Terns were in a poor state with declining colonies from Sussex to Yorkshire and on the North-west coast (Ravenglass was the only protected colony). It was doing better in Wales, but in Scotland it was only found south from the Moray Firth. Some nested on the Isle of Man and in Ireland there were colonies along the east coast and from Clare to Donegal. In 1984 there were 257 pairs counted, less than 15 years earlier, but the total was down to 174 pairs by 1995. In Scotland the birds have expanded their range northwards to Caithness (even Orkney) and there are several colonies now along the West coast and in the Outer Hebrides – probably 300 pairs in all. There may be over 50 pairs in Isle of Man and in Wales (Flintshire and, possibly, Llyn peninsula) but the stronghold is undoubtedly England – and particularly East Anglia. Well over three-quarters of the population (about 2,000) breed from Lincolnshire to Hampshire. Modern protection, the provision of shelters for the nestlings, predator control and fencing can be very successful leading to excellent productivity from concentrated colonies. *Given proper protection and encouragement these birds can do well.*

Hannon, C., Berrow, S.D. & Newton, S.F. 1997 *Irish Birds*: 6, 1-22.

BLACK TERN *Chlidonas niger*

This was a common breeder in the 18th century in eastern marshlands but declined quickly so that by 1850 it was only sporadic. A small colony alleged to have nested at Pett (Sussex) in 1941 and 1942 is now discounted. Recorded breeding attempts since then are: Ouse Washes (several in 1966 and 1969 & 1975), Lough Erne, Fermanagh (1967 & 1975), Tayside (taking food to young in 1970) and Nottinghamshire (1975). There has been a long-term decline in adjacent countries in Europe. *Should we give up hope?*

GUILLEMOT (Common Guillemot)

Uria aalge

> *Distribution* Britain 212 (-5.8%) Ireland 59 (-20.3%)
> *Numbers breeding*: Britain 1,050,000 Ireland 153,000 adults
> *European status*: 2,000,000 (30% in Britain and Ireland =2)
> *British population trend*: continued increase – unabated?
> *How likely are you to record it?* 3 squares (0.1%) *Ranked* 171= [101=]

These birds, our largest auk, have suffered from sport shooting and egg collecting, from oil pollution and from being caught in fishing nets and yet they are doing very well indeed. The appalling exploitation of the seas resulting in over-fishing have probably increased the biomass of small fish available for the birds by reducing the populations of larger, predatory fish. Victorian sport shooting at the colonies, cliff falls and later oil pollution extinguished the eastern colonies along the Channel coast by 85 years ago. There have been recent reductions in the number of colonies in south-west Britain and, probably, south-west Ireland but the number of breeding birds is increasing. In many areas these increases are very fast – Rathlin Island in the North of Ireland may have 80,000 now where there were only 21,575 in 1969. The main colonies are in Scotland and many of them have also been increasing rapidly – by 10% *per annum* or more and the same goes for the colonies in North-east England. Here the estimate was 15,528 in 1969-70 and 59,244 in 1985-87! The last major sport shooting abroad (at sea) was much reduced 30 years ago; birds are still caught in fishing nets but the toll of oil pollution, which was appalling during the period 1940 to 1970, is not now nearly so important. Even the *Erika* incident, with possibly 50,000 Guillemots dead off North-west France, may only have a long-term effect on 'local' colonies. *Excellent prospects for the future.*

RAZORBILL

Alca torda

Distribution Britain 233 (-3.3%) Ireland 63 (-23.2%)
Numbers breeding: Britain 148,000 Ireland 34,000 adults
European status: 480,000 (19% in Britain and Ireland =2)
British population trend: further increases expected
How likely are you to record it? 3 squares (0.1%) *Ranked* 171=

The distribution of the Razorbill is very like that of the Guillemot and their early history is much the same. However there seems to have been little change in numbers in many areas during the first half of this century. Since then there does seem to have been a general increase (but see below) but counting this species, which normally nests in concealed sites, is very difficult. Some Scottish colonies may have increased by 10% *per annum* or more over periods of 10 years or more. In Ireland the situation is very different. At Rathlin, Antrim, numbers have increased very rapidly – possibly five fold in the last 30 years. At others the birds have been in decline for many years and this has been blamed, by some people, on deaths caused by illegal salmon drift netting. Just possibly this has been caused by global warming and not by direct deaths. Both the Razorbill and Guillemot are coming back to the colonies to claim their breeding sites much earlier (by several weeks) which is possibly an indication that the birds are doing very well. *Again the prospects are very good.*

GREAT AUK *Pinguinus impennis*

Extinct in 1844 (Iceland) the last British record was in 1840 on Stac an Armin, St Kilda and the last Irish bird was off Waterford in 1834.

BLACK GUILLEMOT

Cepphus grylle

> *Distribution* Britain 383 (+27.8%) Ireland 90 (-28.6%)
> *Numbers breeding*: Britain 36,500 Ireland 3,000 adults
> *European status*: 100,000 (20% in Britain and Ireland =2)
> *British population trend:* possible increases
> *How likely are you to record it?* 4 squares (0.1%) *Ranked* 167=

These auks nest in much smaller colonies than the others and feed close by on small fish taken in the shallow waters that they like. In the past the breeding area has fluctuated, back and forth, but has remained largely the same. South of the Moray Firth there are few on the east coast but some bred, many years ago, in Yorkshire. On the west there are numerous recent colonies in South-west Scotland and the scattered colonies along the east coast of Ireland are mostly new and may be increasing. A handful of pairs now breed in Anglesey and there is a good population on the Isle of Man. About three quarters of the British and Irish birds breed in Orkney, Shetland and the Western Isles and the current survey, Seabird 2000, will produce a much better estimate of breeding numbers than we have had before. As has been proved several times in Orkney and Shetland, these birds are vulnerable to oil pollution and recovery takes ten years or more. *These birds may be in an expansion phase at the moment.*

PUFFIN (Atlantic Puffin)

Fratercula arctica

> *Distribution* Britain 151 (-17.1%) Ireland 25 (-19.4%)
> *Numbers breeding*: Britain 898,000 Ireland 41,000 adults
> *European status*: 5,400,000 (9% in Britain and Ireland =3)
> *British population trend*: some declines seem likely

These iconic seabirds breed in rather few large and often remote colonies often on islands. They are very difficult to count as their nests are underground in burrows or within tangled rock falls. It is thus not easy to relate earlier estimates of numbers to the current situation. However it is clear that Hirta (St Kilda), where there are large areas of old burrows no longer occupied, remains the largest colony. Few now nest on Grassholm which lost its peat through erosion more than 100 years ago, and most Welsh colonies are now much reduced. There was a spectacular increase in numbers on the Isle of May (Firth of Forth) from about 10 pairs in 1950 to about 12,000 in 1984 but then adult survival declined and the increase stopped. In Ireland there have been some 10 substantial colonies and several of the southerly ones have declined over the last few decades. Quite recently these birds seemed to be doing well but, possibly, they are now suffering the effects of global warming. As a species with a northerly distribution, the conditions may not suit them and this might be the reason that adult survival has declined and why fledging success at St Kilda went down over the last 25 years. *Still loads of them but fears for the future.*

Harris, M.P., Murray, S. & Wanless, S. 1998 *Bird Study*: **45**, 371-374.
Harris, M.P. *et al.* 1997 *J. Avian Biology*: **28**, 287-295.

PALLAS'S SANDGROUSE *Syrrhaptes paradoxus*

These birds were forced into long distance movements from their breeding grounds on the steppes by late lying snow in several years from 1859 to 1908. The biggest was in 1888 and breeding took place subsequently near Elgin (Scotland) and in Yorkshire and probably in Norfolk and Suffolk.

ROCK DOVE AND FERAL PIGEON

Columba livia

> *Distribution* Britain 2,086 (+39.2%) Ireland 345 (-14.6%)
> *Numbers breeding*: Britain 200,000 ? Ireland 50,000 ?
> *European status*: 5,400,000 (1% in Britain and Ireland =6)
> *British population trend*: stable? (+15% BBS)
> *How likely are you to record it?* 1246 squares (27.7%) *Ranked* 37 [55]

This species of bird varies from a hodge podge of different looking pigeons in city centres, derived from many generations of escaped domestic birds, to the smart, truly wild, double black wing-barred denizens of the far North and West. The mixed birds have been around for hundreds of years but largely ignored by birdwatchers. However many towns and cities have very large populations and they are often a great nuisance to those who wish to keep their properties clean, and a great joy to the tourists in Trafalgar Square! Recently some will have been taken by female Sparrowhawks but the increase of Peregrines, and their spread into towns, may limit the continued growth of the feral populations and even reduce them. *The wild populations are safe but the feral ones might be in for a shock!*

STOCK DOVE (Stock Pigeon)

Columba oenas

> *Distribution* Britain 1,821 (-6.8%) Ireland 363 (-32.7%)
> *Numbers breeding*: Britain 240,000 + Ireland 30,000
> *European status*: 510,000 (53% in Britain and Ireland =1)
> *British population trend*: increasing (+124% CBC, +15% BBS){+140%}
> *How likely are you to record it?* 1226 squares (27.3%) *Ranked* 38 [47]

There was great confusion, even to the end of the 19th century, between Stock Doves, Feral Pigeons and Rock Doves. A century ago the range of this species over England and parts of Wales is likely to have been very much the same as now. However there are none in the Highlands and Islands of Scotland and there is some evidence that it spread rapidly northwards in the last 30 years of the 19th century. The birds are still sparsely distributed through parts of Ireland and here, also, they seem to have colonised and expanded as arable farming altered the living conditions in their favour. A steep decline in numbers, coinciding with the use of chemical seed-dressings in the 1950s, was reversed by the late 1960s and the birds are continuing to do well in arable areas although they have disappeared from many peripheral areas. *Modern agriculture does not yet seem to be a problem for this species.*

WOODPIGEON (Common Wood Pigeon)

Columba palumbus

> *Distribution* Britain 2,510 (-2.3%) Ireland 945 (-1.2%)
> *Numbers breeding*: Britain 2,350,000 Ireland 900,000
> *European status*: 10,000,000 (33% in Britain and Ireland =1)
> *British population trend*: increasing apace (+55% CBC){+201%}
> *How likely are you to record it?* 4012 squares (89.3%) *Ranked* 1 [5=]

This is the largest pigeon and was a bird of woods until it started to breed in isolated trees, hedgerows, villages and towns from about 150 years ago. The birds have spread to the far west of Ireland and are now colonising the islands and in Scotland they had occupied most of the mainland many years ago but, more recently, are to be found breeding on Orkney, Shetland and the Western Isles. Many died through toxic chemical poisoning in the 1950s and 1960s but the population did not suffer noticeably. There is a steady increase in CBC index from a very strong level. This may be because the birds are able to survive well in winter, grazing growing crops, and the arrival of autumn sown oilseed rape may be beneficial. Concerted efforts to reduce population through organised shooting, in February, when the game shooting season has closed, seem to have no effect. *Onwards ever and upward!*

COLLARED DOVE (Eurasian Collared Dove)

Streptopelia decaocto

> *Distribution* Britain 2,210 (+6.7%) Ireland 559 (-2.8%)
> *Numbers breeding*: Britain 200,000 Ireland 30,000 territories
> *European status*: 7,000,000 (4% in Britain and Ireland =6)
> *British population trend*: up and up! (+514% CBC){+782%}
> *How likely are you to record it?* 2167 squares (48.3%) *Ranked* 26 [43]

This grey and noisy small dove is *the* success story of the 20th century. They started to breed in Britain when they reached Norfolk in 1955. They were in Kent and Lincolnshire and had reached Scotland in 1957. Ireland was colonised two years later with birds breeding in Dublin and also, far to the west, in Galway and Wales in 1961. They had been no nearer than Hungary in 1932! By 1970 there may have been as many as 25,000 pairs in Britain and Ireland. Even the CBC index, biased towards

areas colonised early on, increased five fold between 1972 and 1996! The birds now breed over most of the country where there are human settlements but are rather less common in the far west and on islands and high ground. *Definitely here to stay!*

TURTLE DOVE (European Turtle Dove)

 *Streptopelia turtur*

> *Distribution* Britain 940 (-24.9%) Ireland 20 (-35.5%)
> *Numbers breeding*: Britain 75,000 – Ireland 0
> *European status*: 2,200,000 (4% in Britain and Ireland =7)
> *British population trend*: further losses (-62% CBC){-77%}
> *How likely are you to record it?* 436 squares (9.7%) *Ranked 65*

This richly coloured dove, a long distance migrant to West Africa, has always been a bird of England and, to a lesser extent, of Wales. There have been a couple of dozen breeding records in Ireland but none since 1977. In Scotland about 10 breeding records, in the Borders and Lothians, have been recorded. Between 1900 and 1950 or so the birds spread north and west into Lancashire, Yorkshire, Northumberland and west into Wales. They probably were as widespread as they ever were in the 1960s. Since then they have retreated from the North and the West and have been lost from many peripheral areas. The population loss, from the CBC, has been serious and steady. The agricultural change, with reduced weed seeding and early planting of crops, has reduced their chances in Britain but the birds are also shot on migration through France and Iberia. There are fewer breeding attempts made by each territorial pair in recent years. *Unless proper agri-environment measures are introduced there is little chance of reversing this depressing trend.*
UKBAP **MAFF** RSPB & English Nature.

ALEXANDRINE PARAKEET
Psittacula eupatria

Several breeding records in scattered locations of escaped birds – the latest in 1997 near Liverpool.
Ogilvie, M.A. & RBBP 1999 *British Birds*: **92**, 176-182 & 472-476.

RING-NECKED PARAKEET (Rose-ringed Parakeet)
Psittacula krameri

Distribution Britain 63 (+1475%) Ireland 0
Numbers breeding: Britain 1,550 Ireland 0
European status: 3,350 (90% in Britain and Ireland =1)
British population trend: gradual increases and expansion
How likely are you to record it? 15 squares (0.3%) *Ranked* 140=

These exotic birds have been invading Britain for 30 years following escapes from aviaries. Their breeding strongholds in West London and Ramsgate in Kent, with smaller numbers in several other areas – including Wales and Scotland. Careful counts at the known roosting sites indicate that the British population is just over 1,500 compared with about 1,000 in 1986. They are well able to withstand cold weather in the winter provided they can find enough food – and lots of people love to feed their exotic visitors. These birds are seldom recorded breeding but lack of nest sites does not seem to be a problem. *Outlook as bright as their plumage – but is this a* **good thing?**
Pithon, J.A. & Dytham, C. 1999 *Bird Study*: **46**, 112-116.
Ogilvie, M.A. & RBBP 1999 *British Birds*: **92**, 176-182 & 472-476.

MONK PARAKEET *Myiopsitta monachus*

Another parrot with a history of escaped parties in several areas. In 1997 there were ten at a Devon locality where breeding had taken place in earlier years and a pair was observed nest building in Surrey.
Ogilvie, M.A. & RBBP 1999 *British Birds*: **92**, 176-182 & 472-476.

BUDGERIGAR *Melopsittacus undulatus*

This common cagebird is often seen as an escaped singleton in a flock of House Sparrows but they are small birds and much more vulnerable to cold than their bigger cousins. Deliberate attempts to establish free-flying breeding colonies have all, so far failed – even on the Scillies.

Cuckoo chick and Robin.

CUCKOO (Common Cuckoo)

Cuculus canorus

> *Distribution* Britain 2,418 (-4.9%) Ireland 706 (-25.1%)
> *Numbers breeding*: Britain 19,500 – Ireland 4,500
> *European status*: 1,600,000 (2% in Britain and Ireland)
> *British population trend*: declining (+3% CBC, -15% BBS){-12%}
> *How likely are you to record it?* 1612 squares (35.9%) *Ranked* 32 [33]

Breeding Cuckoos depend on the presence of their preferred host species to be able to succeed – Dunnock, Reed Warbler and Meadow Pipit account for more than 80% of 'breeding events'. They seem to have been common throughout Britain and Ireland 100 years ago, save for Orkney and Shetland. The suspicion of a decline in Ireland, particularly the west and offshore islands, was confirmed by the falling number of squares occupied in the second Breeding Atlas. Birds were also lost in western England, West Wales and Scotland – particularly the area between Aberdeen and Elgin. This has been related to a decline in Meadow Pipits and one of their other preferred hosts, Dunnock, is also in decline. The birds are very conspicuous but have large territories so the CBC is not a good scheme for monitoring them. The 15%, significant, decline in numbers recorded on the BBS over five years is of some concern. *Populations certainly fluctuate but prospects do not look good.*

BARN OWL

Tyto alba

> *Distribution* Britain 1,110 (-37.5%) Ireland 185 (-62.9%)
> *Numbers breeding*: Britain 4,400 Ireland 750
> *European status*: 140,000 (4% in Britain and Ireland =5)
> *British population trend:* decline halted, now steady
> *How likely are you to record it?* 32 squares (0.7%) *Ranked* 125=

The silent moth-like flight of the Barn Owl across grassy fields of open marshland is a sight to stir the heart. By 1900 numbers were already falling with shooting for trophies and because of the irrational fear that anything with talons and hooked bill would attack Pheasants. The birds were spreading north in Scotland but doing badly in the south – later they recovered well until declines set in during the 1950s (possibly due to cold winters or pesticides) but recovered again. In Ireland the birds certainly withdrew from peripheral areas and declined from 1950 or so. The birds are very difficult to survey but an early attempt in 1930 came up with a figure of 12,000 pairs for England and Wales. A Hawk and Owl Trust survey suggested 4,400 for Britain, 600-900 for Ireland and 40 in the Channel Islands during 1982-85. A further survey, run with the BTO, seems to show that the decline in Britain had halted by the late 1990s and the birds may be doing quite well in several areas. These typically have rough grassland and good nest sites. However there are possibly problems with rodenticides, road casualties and, if they happen, bad winters. A recent survey in Ireland (1996/97) suggested that the decline is not yet halted there – only 115 nests found! The birds cannot survive well in cold, wet and windy weather. This is a species which may

be helped by global warming. *Glimmers of hope and lots of human help with nest boxes.*

Berridge, D.J. 1997 *Irish Birds*: 6, 114.
Toms, M. 1999 *BTO News*: 223, 10-12.

EAGLE OWL *Bubo bubo*

This is not a native bird but escaped pairs have successfully bred in Northern England (1997) and Moray and Nairn (1984 & 1985). In their natural range in Northern Europe Eagle Owls kill many other owls from within their extensive territories and many ornithologists think the escaped birds should be returned to captivity. However they would, of course, seek full protection for natural colonists!

Ogilvie, M.A. & RBBP 1999 *British Birds*: 92, 176-182 & 472-476.

SNOWY OWL *Nyctea scandiaca*

This huge, white arctic owl has bred on several occasions on Fetlar, Shetland, since 1967 but there was no male in the breeding season from 1975 – two females remained, waiting and waiting, until 1993.

LITTLE OWL

Athene   noctua

> *Distribution* Britain 1,228 (-11.0%) Ireland 0
> *Numbers breeding*: Britain 9,000 Ireland 0
> *European status*: 250,000 (4% in Britain and Ireland =8)
> *British population trend*: some declines likely (+8% CBC){+17%}
> *How likely are you to record it?* 195 squares (4.3%) *Ranked 87*

Britain had no native small owls and many attempts were made to introduce Little Owls from more than 150 years ago and two were ultimately successful in Kent (1874-80) and Northamptonshire (1888-

90). The birds quickly spread, particularly from the second site, and were aided by further releases. By 1920 they were nearly all over Britain south of the Humber. To the north the gradual expansion took longer. Some birds were reported from Scotland in 1925 but breeding was not recorded until 1958 and the birds remain scarce and almost all in the south-east corner. No breeding has been reported from Ireland. The birds may have suffered a decline from pesticide poisoning in the 1950s and 1960s. The second Breeding Atlas records significant losses from Devon, Cornwall and Lincolnshire whilst the CBC showed an increase in the late 1970s and a fall back to the previous level in the late 1980s. BBS losses of 21% (not significant) in five years are worrying although global warming might help. *Further losses seem likely.*

TAWNY OWL

Strix aluco

> *Distribution* Britain 2,054 (-10.9%) Ireland 0
> *Numbers breeding:* Britain 20,000 – Ireland 0
> *European status:* 470,000 (4% in Britain and Ireland =6)
> *British population trend:* worrying declines (-33% CBC){-30%}
> *How likely are you to record it?* 173 squares (3.9%) *Ranked 93*

Over most of mainland Britain this is the most familiar owl whose night-time hooting is more often heard than the bird is seen. Originally a woodland owl, the bird is now found in fairly open farmland provided there are small copses, or even hedgerow trees, available for them. They have spread north to Caithness in Scotland and recently colonised some of the inner islands of the Hebrides – not recorded breeding on the Western Isles, Orkney or Shetland. Increases on high mainland sites are due to aforestation in all areas. They do not breed in the Scilly Islands or Channel Isles, Isle of Man or Ireland (no records of wild birds) and is sporadic on the Isle of Wight. The recent reduction in occupied squares between the two Atlases, mainly in Scotland and the south-west peninsula and the decline in the CBC indicate that there may be a problem. Some people think this is associated with direct and indirect effects of agricultural chemicals on the bird's small mammal diet. The new generation rodenticides, following the development of rat resistance to warfarin, are much more toxic to birds. *Cause for considerable concern.*

LONG-EARED OWL

Asio otus

> *Distribution* Britain 445 (-24.6%) Ireland 230 (-33.9%)
> *Numbers breeding*: Britain 2,350 Ireland 2,350
> *European status*: 205,000 (2% in Britain and Ireland)
> *British population trend*: contracting and declining
> *How likely are you to record it?* 5 squares (0.1%) *Ranked* 162= [101=]

The Long-eared Owl is very difficult to find and was not widely recorded until 150 years ago when there seems to have been a widespread increase and range extension in Britain. This may have been due to lack of competition from the bigger, and declining, Tawny Owl and the widespread planting of conifers (where it likes to nest). It is the commonest owl in Ireland, where there are no Tawnies, with 230 occupied squares in the latest Breeding Atlas compared with 185 for Barn Owl – much more conspicuous. There were many less records in the most recent Atlas because of changes in methodology, not necessarily fewer birds. However in Britain, except in the remoter areas where there are isolated conifers, they began to lose out as the Tawny Owl population increased. There are now large areas of England and Wales without recent breeding records – particularly south and west of a line from the Mersey through London. *Still on the wane.*

SHORT-EARED OWL

Asio flammeus

> *Distribution* Britain 679 (-15.2%) Ireland 11 (+1000%)
> *Numbers breeding*: Britain 2,250 Ireland 0
> *European status*: 17,000 (13% in Britain and Ireland =5)
> *British population trend*: sporadic but declining overall
> *How likely are you to record it?* 36 squares (0.8%) *Ranked* 124

The Short-eared Owl is the least nocturnal of our owls and is regularly seen quartering rough grassland, moorland and saltmarsh. It is very dependant on the vole populations and may breed densely, and successfully, when their prey numbers are high. In the 19th century Scotland, Northern England and East Anglia had regular breeding populations. The few breeding records in Ireland have mostly been in the last 30 years but they are unlikely to do well since voles are absent from most of the island. Breeding is often associated with newly planted

conifers and this may have led to an increase in many areas and the development of a thriving populations in Wales and the Isle of Man. The second Breeding Atlas recorded many lost squares. This may be due to the lack of new conifer planting. Numbers breeding in the country may fluctuate quite widely due to gluts or dearths of voles elsewhere. *Further declines likely.*

NIGHTJAR (European Nightjar)
Caprimulgus europaeus

> *Distribution* Britain 274 (-51.2%) Ireland 11 (-88.2%)
> *Numbers breeding*: Britain 3,400 Ireland <30
> *European status*: 240,000 (1% in Britain and Ireland)
> *British population trend:* decline halted, real recent increases
> *How likely are you to record it?* 2 squares (0.0%) *Ranked* 183=

A hundred years ago the churring of breeding Nightjars could be heard just about throughout the land – not in the Western Isles or Orkney or Shetland. However by the 1930s a general decline seems to have set in through much of Britain and, about 20 years later, in Ireland. The loss from Ireland is almost complete with only a handful of records in recent years, but clear-felling starting now of 30 and 40 year-old plantations could lead to a temporary improvement. The birds have dwindled over most of Scotland with no more than 100 pairs, at most, in 1981. Then there may have been almost 2,000 churring males in the whole of England

and Wales – most in South-east England and Breckland. The losses looked awful when the second Breeding Atlas was produced but already there was a considerable recovery in Breckland – 90 pairs in 1974, 168 in 1981 to 300 in 1989 probably by taking advantage of the regular pattern of harvesting the plantations. By 1997 there were well over 500! A national survey in 1981 estimated the British population at about 2,100 males and a repeat in 1992 came to 3,400. Scotland did little better than marking time but in Wales the count increased by 230%! Global warming may also be important for a species with such a late arrival date and restricted breeding season in which, potentially, to fit two broods. *A **real** recovery seems to be taking place.*

Morris, A. *et al.* 1994 *Bird Study*: 41, 181-191.
UKBAP **Forestry Commission** RSPB.

SWIFT (Common Swift)

Apus apus

> *Distribution* Britain 2,215 (+0.7%) Ireland 743 (-14.4%)
> *Numbers breeding*: Britain 80,000 Ireland 20,000
> *European status*: 4,300,000 (2% in Britain and Ireland = 8=)
> *British population trend:* recent steep decline (-13% BBS)
> *How likely are you to record it?* 1797 squares (40.0%) *Ranked* 31 [41=]

Swifts are long distance migrants and are only in our area for about three months (May to July) to breed mostly in the roofs of buildings. Only a handful nest in natural sites like cliff crevices, woodpecker holes and even House Martin nests! They were widespread and common 100 years ago breeding everywhere except Orkney, Shetland, the Hebrides and North-west Scotland. Declines had been, and were later, reported from the polluted city centre areas of London, the Midlands and the North. However, as 'Clean Air Acts' were introduced they were able to return. In Ireland they expanded into parts of the far west about 60 years ago but they may have declined since. The Second Atlas showed many (14.4%) of the occupied squares in Ireland had none recorded and Britain showed no change. On the BBS they declined by 13% (significant) in five years. Many are excluded from their breeding holes by demolition and restoration work. One survey of a small part of Northamptonshire showed a staggering decline of almost 80% between 1978 and 1998! This was very closely correlated with re-roofing work. *Concern for Swifts* (see page 76) has been formed to try to alert owners to their guests and tell house owners, councils and house builders how to accommodate their Swifts. *The birds need help or they will largely be lost!*

KINGFISHER (Common Kingfisher)

Alcedo atthis

> *Distribution* Britain 1,224 (-6.2%) Ireland 306 (-40.4%)
> *Numbers breeding*: Britain 4,400 Ireland 1,700
> *European status*: 52,500 (12% in Britain and Ireland =2)
> *British population trend:* probably increasing (+22% WBS){+20%}
> *How likely are you to record it?* 93 squares (2.1%) *Ranked* 106 [83]

The bright, electric flash of a passing Kingfisher is enough to brighten anyone's day. However the beauty and rarity of the bird was the reason for much destruction in the 19th century when there was a thriving trade in stuffed specimens, birds for the adornment of lady's hats and feathers for fly tying. The first protection laws led to some evidence of an increase at the turn of the century but the birds could not come back to areas where river pollution had killed off their food. They can breed very productively to make up for losses in severe winters. The big freeze of 1963 seems to have knocked them back and there are now about 50 10-km squares newly occupied north of the Forth/Clyde in the second Breeding Atlas. In Ireland they were found in many fewer squares during fieldwork for the second than for the first Breeding Atlas. However, apart from local losses, there have been gains in the industrial Pennines probably due to cleaning up pollution. Unsympathetic drainage work and poor water quality is bad for them. Global warming, modern ideas about drainage and better control of pollution could turn the population round in the future. *Increases in prospect provided there are warm winters.*

BEE-EATER (European Bee-eater)

😊

Merops apiaster

A pair tried to breed in 1920 in Midlothian but the female died. Three pairs settled in a sand pit, using Sand Martin holes, in Sussex in 1955 and two raised young. Birds overshooting in spring seem to be increasing and further, sporadic, breeding might occur.

HOOPOE

😊

Upupa epops

This is a spectacular and sporadic breeder in Britain with about 20 attempts in the last 200 years – mostly in fine summers when spring migrants have overshot. The best, recent, year was 1977 when four pairs were proved but, since then, a Welsh pair in 1996 is the only proved record. A pair in Greater Manchester in 1997 may have bred. Pairs have summered in Ireland (latest 1982 Kilkenny) but they have never been proved to breed.

WRYNECK (Eurasian Wryneck)

 😞😞

Jynx torquilla

> *Distribution* Britain 6 (-87.5%) Ireland 0
> *Numbers breeding*: Britain 5 RBBP Ireland 0
> *European status*: 380,000 (0% in Britain and Ireland)
> *British population trend*: – no real population!

It is difficult for young birdwatchers to realise that this species was so common, less than 100 years ago, that it was known as the Cuckoo's mate. The loud ringing song was as familiar as that of the Cuckoo and both migrants arrived at roughly the same time. It was actually the commonest woodpecker in many areas! But the writing was on the wall by 1900 as it was no longer breeding in Northern England or in North Wales. The decline accelerated and by 1930 it was very scarce in the Welsh borders and over much of the Midlands. By 1950 there were birds still in Norfolk, Suffolk and Kent and along the Thames Valley but by 1965 the population was down to 25-30 pairs. From 1951 there had been calling birds in Scotland every few years and in 1969 three pairs were proved to breed on Speyside (clearly from the Scandinavian breeding population). Breeding is now sporadic with the average maximum estimate (RBBP) about five pairs over the last ten years and only two proved in the five years 1993-97 – the latest a Scottish one in 1997. The reasons are clearly

related to the species' withdrawal from the north of its range in Western Europe, but a good case has also been made for it to be related to the loss of old pasture with the ants that the bird relies on for food. *These birds might have gone for ever.*
UKBAP **Scottish Natural Heritage** RSPB.

GREEN WOODPECKER

Picus viridis

Distribution Britain 1,555 (-4.1%) Ireland 0
Numbers breeding: Britain 15,000 ++ Ireland 0
European status: 670,000 (2% in Britain and Ireland =9)
British population trend: increasing (+15% CBC){+92%}
How likely are you to record it? 1197 squares (26.7%) *Ranked 39*

The loud, cheerful, ringing call of the Yaffle is a characteristic sound of open woodland, parkland and much farmland where there is a scattering of mature trees. The birds do not find much of their food from old wood, preferring to feed on ants – even clearing away snow to get at them. From being a breeding bird throughout England and Wales and into Scotland as far as Perth it retreated in the first half of the 19th century. None bred in Scotland after 1850 and it became rare in northern England – possibly related to trophy shooting. In the 20th century it re-colonised some northern counties – in the east by 1935 and the west a bit later. Breeding was proved in Scotland in 1951 (suspected rather earlier) and the spread through the country was pretty rapid. The lowlands were well colonised by the time of the first Breeding Atlas and they had reached the Black Isle, the Great Glen and Kintyre by the second with later records from as far north and west as Mull. They have never bred in Ireland. All is not gain. The birds have also been lost from areas of Cornwall, West

Wales and much of Dumfries. They can be badly affected by cold winter weather and so may benefit from global warming. Recent population index information from the CBC and BBS is encouraging. *Prospects seem good but western decline a bit of a worry.*

GREAT SPOTTED WOODPECKER

Dendrocopus major

> *Distribution* Britain 1,959 (-4.4%) Ireland 0
> *Numbers breeding*: Britain 27,500 Ireland 0
> *European status*: 3,700,000 (1% in Britain and Ireland)
> *British population trend:* increasing strongly (+185% CBC){+161%}
> *How likely are you to record it?* 1265 squares (28.2%) *Ranked 36*

About two hundred years ago something happened to cause this woodpecker to retract from the periphery of its breeding range. The birds were lost from Scotland, where the Caledonian pine forests had been their stronghold, and from most of northern England and parts of Wales. They became uncommon even in some southern counties too. It has been suggested that intensive woodland management with dead wood being taken for fuel might have, at least in part, been the cause. None, of course, breed in Ireland although bones found indicate they were around in early Christian times – possible lost with the final forest clearances in the 17th century. They started to make good their losses in the late 1860s and re-colonised Scotland in 1887 (Berwickshire). The birds spread quickly with the Caledonian pine re-occupied before 1900 and the very north of Caithness, Skye and Mull before the end of the 1960s. This is a bird which seems to have thrived in the main parts of its range recently. Additional dead wood caused by Dutch Elm disease and the use of garden bird feeders may have helped a lot and we have not had any very cold and freezing winters for some years. The big increase in CBC indices (179% on farmland and 89% on woodland between 1972 and 1996) and significant 36% increase on the BBS sites in five years are very dynamic. *A species on the up and up.*

LESSER SPOTTED WOODPECKER

Dendrocopus minor

> *Distribution* Britain 790 (-11.1%) Ireland 0
> *Numbers breeding*: Britain 4,500 Ireland 0
> *European status*: 210,000 (2% in Britain and Ireland=9)
> *British population trend:* peaked early 1980s, crashed since (-76%CBC){-42%}
> *How likely are you to record it?* 45 squares (1.0%) *Ranked 116=*

For many people our smallest woodpecker is a bird of mystery. Many of them, living in big deciduous trees, seldom come lower than 10 metres and are often very difficult to see. In the 19th century these birds seem to have bred throughout England and Wales except for the very north and the extreme west. Some losses were recorded in areas where the habitat was cleared – mostly through felling of orchards, for instance, when the cider apple trees were cleared in Hereford and Somerset. There are no records from Ireland and only very tentative possible breeding records from Scotland. The population is hard hit by cold winter weather but seems to recover fairly well. The spread of Dutch Elm disease with the consequent extra dead wood helped. However the effect did not last – possibly because the thinner dead wood was quickly removed from the trees by gales – and the high level of the CBC from 1976 to 1984 dropped quickly. The loss is more than 80%, very worrying as it does not coincide with cold winters. The losses documented in the second Breeding Atlas are particularly concentrated in the West Country and West Midlands. There was a slight extension in breeding range into Northumbria. *Recent trends almost all point to the species being in real trouble.*

WOODLARK (Wood Lark)

Lullula arborea

> *Distribution* Britain 73 (-62.8%) Ireland 0
> *Numbers breeding*: Britain 1,552 RBBP Ireland 0
> *European status*: 1,400,000 (0% in Britain and Ireland)
> *British population trend:* huge increase: + 600% in 12 years
> *How likely are you to record it?* 37 squares (0.8%) *Ranked* 123

The lovely song of the Woodlark has probably never been very commonly heard in the country but it was certainly widespread 150 years ago. Birds

bred in all but the northernmost counties of England, through most of Wales and along the east and south coasts of Ireland. By 100 years ago they were mainly breeding near or south of the line from Gloucester to the Wash with the Irish birds lost from their last site in Wicklow – there were two later breeding records in Wexford (1905) and Cork (1954). In Britain the decline carried on for another 20 years but then they began to increase again and re-colonised many areas North to Nottinghamshire, Lincolnshire and Yorkshire. However by 1955, and particularly after the cold winter of 1962/63, they declined again to as little as 100 pairs. By the time of the first Breeding Atlas this had at least doubled (200-450 pairs) and then dropped again in 1975 to 160-180 pairs. Recovery was again quick with 400-430 pairs in 1981 but this was halved following cold winter weather in 1981/82. For the second Breeding Atlas 350 breeding pairs were estimated and by 1993 600 was the figure. By 1999 there were almost this many breeding in the Brecks alone! *Excellent prospects for further increase and range expansion.*
Sitters, H.P. *at al.* 1996 *Bird Study*: 43, 172-187.
UKBAP **Forestry Commission** RSPB.

SKYLARK (Sky Lark)

Alauda arvensis

> *Distribution* Britain 2,729 (-1.6%) Ireland 927 (-6.4%)
> *Numbers breeding*: Britain 1,046,000 Ireland 570,000
> *European status*: 30,000,000 (8% in Britain and Ireland =4)
> *British population trend:* recent sharp declines (-60% CBC){-52%}
> *How likely are you to record it?* 3068 squares (68.3%) *Ranked* 13 [20]

This is one species that earlier birdwatchers would never have believed could cause real concern because of widespread declines. Universally common in open areas from coastal marshes almost to the mountain tops they were caught and caged, killed and eaten at home or exported, in huge numbers, for food, mostly to France. There was no suggestion of a problem at the time of the first Breeding Atlas but between this and the second Atlas there was a 60% decline on the CBC – 75% on farmland. This is still a common and very widespread bird, but even recent surveys in the remote peatlands of the Flow Country of Caithness and Sutherland (undertaken in 1988, 1991 and 1995) showed tremendous losses. Skylark registrations fell from 1235 to 565 and finally 238 – a staggering decline of 81% over seven years. This is probably a winter problem on farmland further south. Detailed work on the species indicates that they may be breeding less often on arable land as the crops are too dense for them to nest. BTO data from NRS shows that recent breeding attempts are

actually becoming more successful. However in the winter the lack of stubbles, through autumn planting, and weed seeds, through efficient herbicide use, are probably affecting survival badly. *Until the CAP is revised and helps the species, more losses are almost inevitable.*
Browne, S., Vickery, J., & Chamberlain, D., 2000 *Bird Study*: 47,52-65.
Chamberlain, D.E. & Crick, H.Q.P. 1999 *Ibis*: **141**, 38-51.
Hancock, M. & Avery, M. 1998 *Scottish Birds*: **19**, 195-205.
UKBAP MAFF RSPB.

SHORELARK (Horned Lark)

(° °)   *Eremophila alpestris*

This attractive species flirted with breeding in Scotland, high in the mountains, for a few years and there is one recent record (1997) of a pair hinting of breeding. Between 1972 and 1977 singing birds or pairs were found in the same locality and breeding was proved in 1977 when there was another singing bird at a second site.

SAND MARTIN

Riparia riparia

> *Distribution* Britain 1,559 (-23.7%) Ireland 595 (-29.3%)
> *Numbers breeding*: Britain 160,000 Ireland 100,000
> *European status*: 2,400,000 (11% in Britain and Ireland =2)
> *British population trend*: recovering after slumps
> *How likely are you to record it?* 192 squares (4.3%) *Ranked* 89 [62]

These little brown swallows need vertical banks to nest in – normally digging their own holes in soft substrates but sometimes using existing holes (usually drains). Eroded banks beside streams or rivers, on the shores of lakes or even sea cliffs are typical natural sites. Colonies of hundreds, even thousands, of pairs can settle at sand and gravel pits where the workers were, and are, often very possessive of 'their' birds. The ephemeral nature of the natural sites means that the birds are ideally equipped to exploit newly excavated sites. In areas where peat digging takes place, particularly in Ireland, the birds may use these for nesting and piles of washed sand, sawdust end even foundation excavations and rabbit holes have been used! In the last 15 years specially constructed breeding sites have been built for the birds on various reserves.

The population of Sand Martins in Britain and Ireland has undoubtedly shifted from natural to artificial sites. Sporadic breeding on the Western Isles, Shetland and Orkney seems to have stopped about 80 years ago. By 1968 a peak was reached and then, following a severe

drought in the winter quarters and heightened mortality, the population dropped by two thirds or more over two years. There was a reasonable recovery but a further drop in 1984, possibly to a tenth of the peak population, again caused by Sahel drought. There has been another recovery but the BBS results (-21% not significant) may indicate that this has ended. In any event this species is clearly at risk from climatic disaster far from our shores. *Outlook may be better in Britain and Ireland but uncertain in winter.*
Bryant, D.M. & Jones, G. 1995 *Bird Study*: **42**, 57-65.

SWALLOW (Barn Swallow)

Hirundo rustica

> *Distribution* Britain 2,626 (+1.1%) Ireland 982 (-0.4%)
> *Numbers breeding*: Britain 570,000 Ireland 250,000
> *European status*: 15,000,000 (5% in Britain and Ireland = 6)
> *British population trend*: some declines but CBC up 20%{-16%}
> *How likely are you to record it?* 3111 squares (69.3%) *Ranked* 11 [4]

This bird is the symbol of summer and was a familiar summer visitor all over Britain and Ireland save for the very west of Ireland and the Outer Hebrides 100 years ago. Nesting was not common on Orkney and was sporadic on Shetland. The birds must have learnt to use the typical man-made nesting sites a few hundred years ago since it was earlier called the 'Chimney Swallow'. The birds certainly colonised western Ireland (including some of the islands) and became better established in the distant parts of Scotland by the time of the first Breeding Atlas. There were

further gains on the Outer Hebrides and Orkney, particularly, logged in the second Atlas. However there are many reports from the Home Counties of Swallows being lost as a breeding species. Large numbers of farms have lost their livestock and the consequent lack of insects may be a problem for Swallow populations. The birds are not easy to census accurately for the CBC and many younger birdwatchers have no memory of just how common Swallows used to be in rural areas. Recent thinking that the number of breeding attempts by each breeding pair may be very important in determining the long-term health of the species may apply to this species – although NRS results do show earlier and more successful nesting from about 1985 to 1996. *Still serious concerns about a species apparently coping well with modern conditions.*

HOUSE MARTIN

Delichon urbica

Distribution Britain 2,393 (-1.3%) Ireland 810 (-9.4%)
Numbers breeding: Britain 375,000 — Ireland 105,000
European status: 12,000,000 (4% in Britain and Ireland =9)
British population trend: apparently stable but some bad losses{-1%}
How likely are you to record it? 1518 squares (33.8%) *Ranked* 33 [29=]

There are still natural breeding House Martin nests on coastal cliffs, particularly in Scotland and Ireland, and a very few inland (sometimes in quarries). However most nest on buildings and have done for many years. The range of the birds seems not to have changed much with sporadic breeding in the Shetland and the Western Isles and sparse nesting in Orkney, the Inner Hebrides, North-west Scotland and Western Ireland. The air pollution in industrial areas and city centres forced the birds out but they returned, to a certain extent, after the passing of the Clean Air Act in 1956. The long term study in Scotland does not show consistent population changes but declines, and total losses, are reported from the Home Counties. Several rural towns with dozens of pairs 20 years ago are now reporting no more than four or five pairs. This has been related to loss of livestock from local farms, increased use of pesticides and the loss of muddy puddles. The latter may be very important as the birds have to use liquid mud, which will stick well, within 200 metres of the nest site or it will become too dry in flight. There is some evidence that the birds are arriving later and departing earlier from their nesting sites in South-east England and this may be driving a local decline. *There may be a serious, local, problem but this has yet to be proved. Let's hope the birds come through OK!*

TAWNY PIPIT *Anthus campestris*

Breeding records earlier the last century in Sussex have now been rejected.

TREE PIPIT

Anthus   trivialis

> *Distribution* Britain 1,524 (-15.0%) Ireland 5 (0%)
> *Numbers breeding*: Britain 120,000 – Ireland <1
> *European status*: 17,000,000 (1% in Britain and Ireland)
> *British population trend*: declining (-56% CBC but up 36% BBS){-62%}
> *How likely are you to record it?* 253 squares (5.6%) *Ranked 82*

There are problems with the old records of this species as there was confusion with the Meadow Pipit. However it does seem that there was considerable spread into the north and west to include all mainland Scotland, save parts of Caithness and Grampian, and even some of the Inner Hebrides in the hundred years before the first Breeding Atlas. Over the same period they also spread west into most of Wales and parts of the West Country. The birds now breed regularly in small numbers on the Isle of Man and may be a very rare and overlooked breeding bird in Ireland. This species is able to take advantage of short-term changes to the habitat in exploiting newly formed clearings in forests. It has certainly done well in exploiting the early stages of forestry plantations. However the CBC changes recorded for 1972-96 of a decline of 56% is a cause for considerable concern. The losses recorded in the second Breeding Atlas were particularly concentrated in South-east England. However an increase of 36% in the five year BBS index (1994-98) may show this is misplaced. *The losses suffered recently do not seem to be simply a part of the normal fluctuations of this species.*

MEADOW PIPIT

Anthus pratensis

> *Distribution* Britain 2,539 (-3.2%) Ireland 945 (-5.5%)
> *Numbers breeding*: Britain 1,900,000 Ireland 900,000
> *European status*: 9,000,000 (32% in Britain and Ireland =1)
> *British population trend*: declines after huge increases (-28% CBC){-20%}
> *How likely are you to record it?* 1356 squares (30.2%) *Ranked* 34 [16]

This is one of the species that bred 'everywhere' in the 19th century but, even then, they were not breeding in lowland arable farmland. They have always had their stronghold in upland rough grassland, heathland, young forestry, uncultivated areas and salt marsh. In Scotland and Ireland they breed commonly on the remotest islands of the north and west but the second Breeding Atlas logged losses from the south-east half of Ireland and from much of lowland England. These losses may, at least in part, be to do with the intensification of agriculture and consequent loss of waste and rough grazing areas. Of considerable surprise were the results from breeding season surveys of 19 remote and unaltered plots in the Flow Country of Caithness and Sutherland, between 1988 and 1995, which showed an overall decline of 46% in registrations (see Skylark page 216). The NRS shows a recent increase in egg stage nest failures but the young stage seems to have become more successful over the last 25 years. *Still very common but we should be rather worried.*
Hancock, M. & Avery, M. 1998 *Scottish Birds*: **19**, 195-205.

ROCK PIPIT

Anthus petrosus

> *Distribution* Britain 654 (-11.0%) Ireland 259 (-9.8%)
> *Numbers breeding*: Britain 34,000 Ireland 12,500
> *European status*: 400,000 (11% in Britain and Ireland =2)
> *British population trend*: slight decline?
> *How likely are you to record it?* 31 squares (0.7%) *Ranked* 128 [77=]

The Rock Pipit has always been a coastal species breeding on just about all rocky shores in Britain and Ireland. In some areas, like Mull, Skye and the Western Isles, it seems regularly to breed a little inland. It is missing from Lincolnshire to the Thames and has only recently returned to the cliffs in Kent. Numbers are low for much of the sandy, low, coastline on the east coast of Ireland, north-west England and the coasts of the Channel and the North Sea. It is from these areas that most of the losses between the two Breeding Atlases were recorded. However there are no widespread census results available from core areas of high density where any population trends could be deduced. Most birds seem to stay in or near their breeding area during the winter. *A hint of a problem is apparent.*

YELLOW WAGTAIL

Motacilla flava

> *Distribution* Britain 1,047 (-9.4%) Ireland 3 (-25.0%)
> *Numbers breeding*: Britain 50,000 Ireland 0
> *European status*: 4,500,000 (1% in Britain and Ireland)
> *British population trend*: widespread declines (-25% CBC){-13%}
> *How likely are you to record it?* 319 squares (7.1%) *Ranked* 70

The race of Yellow Wagtail commonly breeding in Britain is *flavissima* and a few of these breed in Norway, the Netherlands and Northern France. A few of the nominate race *flava* breed here and there are sometimes birds showing the characteristics of other subspecies. At the turn of the century most bred south of the line from the Tees to Morecambe Bay, east of the mountains in Wales and from Dorset eastwards. There were small populations north to the Highlands of Scotland. In Ireland colonies round Lough Neagh had become extinct by 1944 and round the Galway and Mayo Loughs (Corrib, Mask and Carra) by about 80 years ago. In Ireland there have been more than 30 possible breeding attempts, mostly near the East Coast (especially Co. Wicklow marshes), since 1950 with some Blue-headed (*flava*) birds involved. In Scotland the birds retreated as regular breeders from many of their former nesting areas with Ayr and Lanark remaining as the stronghold and only a few pairs or sporadic nesting elsewhere. The second Breeding Atlas showed losses in Wales, southern and northern England and the declines in the CBC, WBS and, most recently, 30% in the five years 1994-98 on the BBS are of great concern. The damp pastures and marshy areas where it used to breed are becoming rarer and rarer. They often nest in tangled crops like vining peas or potatoes, which are harvested too early for the birds to complete even one breeding attempt. *This species is in need of our help.*

GREY WAGTAIL

Motacilla cinerea

> *Distribution* Britain 1,979 (+7.3%) Ireland 816 (-8.6%)
> *Numbers breeding*: Britain 34,000 Ireland 22,000
> *European status*: 720,000 (8% in Britain and Ireland =3)
> *British population trend*: declining in some areas (-34% CBC){-39%}
> *How likely are you to record it?* 271 squares (6.0%) *Ranked* 77 [41=]

These elegant and attractive birds invariably nest by swift running water and, in lowland areas, are almost always found by sluices, weirs, canal locks, etc. They used to breed only in Scotland, northern England and Ireland but by 1900 had colonised Wales and South-west England and had started to spread into the Midlands and along the South Coast to Kent to exploit the available man-made sites. This expansion gradually continued to Essex and Leicestershire in 1951 for instance. Between the two Breeding Atlases many new sites in eastern Britain were colonised. The birds mostly leave their breeding area for the winter and are severely affected by cold winters when the population is often halved. Such setbacks are soon recovered after a few mild winters. The BTO medium alert is because of the 34% decline in population recorded on both the CBC and WBS but it might have been high at the beginning of the period – following a succession of mild winters. *Further increases in lowland areas are to be expected.*

PIED WAGTAIL

Motacilla alba

> *Distribution* Britain 2,669 (+0.3%) Ireland 966 (-1.6%)
> *Numbers breeding*: Britain 300,000 Ireland 130,000
> *European status*: 9,00,0000 (5% in Britain and Ireland = 5=)
> *British population trend*: stable (+ 5% CBC){+39%}
> *How likely are you to record it?* 2020 squares (45.0%) *Ranked* 29 [23=]

The Pied Wagtail is the British and Irish race (*yarelli*) of the widespread species correctly known as the White Wagtail (*alba*) – this race does sometimes nest as pairs or hybrids usually on the South or East coast of England or in the north of Scotland. The species was well distributed 100 years ago. In Ireland only a small part of the very far west was without them – and this and the far islands were occupied by 1954. In Scotland the birds gradually made good their lost ground and started to

breed regularly on the Western Isles (1930s) and Orkney (1970s) and sparsely on Shetland (1970s). There is some evidence that the birds are not as common as they used to be in the areas where the farming has become wholly arable. Population levels from the WBS declined by 39% but the BBS showed a significant increase of 13% in five years (1994-98). The birds are vulnerable to cold winter weather but come back quickly if the subsequent years are mild. *Populations in the lowlands of England may be threatened but, overall, the species seems to be in good heart.*

WAXWING (Bohemian Waxwing)

Bombycilla garrulus

Birds sometimes remain in Scotland until well into the summer after invasions and may even sing and display but there has been no serious suggestion of breeding.

DIPPER (White-throated Dipper)

Cinclus cinclus

> *Distribution* Britain 1,309 (-8.7%) Ireland 429 (-30.4%)
> *Numbers breeding*: Britain 14,000 Ireland 3,500
> *European status*: 180,000 (10% in Britain and Ireland =3)
> *British population trend*: pretty well stable (- 10% WBS){-6%}
> *How likely are you to record it?* 103 squares (2.3%) *Ranked* 102 [69]

We have two races in Britain – *gularis* over most of Britain and *hibernicus* over Ireland and North-west Scotland. The species' distribution is determined by the bird's requirement for reasonably fast flowing streams running over rocky or gravelly beds. The bird's distribution has not changed much although there were periods of regular (now sporadic) breeding on the Isle of Man and Orkney: the birds regularly breed on the Western Isles. Local losses have been recorded from industrially polluted areas – and re-colonisation following clean-up. There are widespread reports of birds becoming scarcer or being lost from rivers and streams subject to acidification which may be caused by plantations of conifers. This reduces the insect and fish life available for the birds. There were many more losses than gains recorded between the two Breeding Bird Atlases and these may indicate a real retreat from marginal areas for the birds. Some of these, in Western England, have only recently been colonised. *There are probably more about to be lost but the species should not be threatened in our area.*

WREN (Winter Wren)

Troglodytes troglodytes

> *Distribution* Britain 2,747 (-0.3%) Ireland 987 (-0.5%)
> *Numbers breeding*: Britain 7,100,000 + Ireland 2,800,000
> *European status*: 22,000,000 (41% in Britain and Ireland =1)
> *British population trend*: fluctuates, basically stable (+4% CBC){+69%}
> *How likely are you to record it?* 3834 squares (85.4%) *Ranked* 4 [1]

These tiny birds are to be found all over Britain and Ireland – even on most of the remotest islands. Several island races have been named and recognised and the birds on St Kilda (*hirtensis*) were shamefully exploited by collectors in the 19th century – but no permanent harm was done. Breeding on some small islands is sporadic but there seems to have been no detectable long-term trend to the range (ubiquitous) or population (the commonest bird in many years). Locally the planting of conifers on open moorland will lead to increase and the clearing of woodland for farmland will cause declines. Short-term fluctuations in population after cold winters, and their recovery after mild ones, are well documented with the worst declines, such as after 1962/63, sometimes reaching 80% or more. The only significant changes recorded from the NRS is an increase in clutch and brood size. *All seems well for a popular bird.*

DUNNOCK (Hedge Accentor)

Prunella modularis

> *Distribution* Britain 2,511 (-3.2%) Ireland 947 (-2.3%)
> *Numbers breeding*: Britain 2,000,000 Ireland 810,000
> *European status*: 11,000,000 (26% in Britain and Ireland =1)
> *British population trend*: widespread declines (-31% CBC){-21%}
> *How likely are you to record it?* 3197 squares (71.2%) *Ranked 9* [14]

A familiar and common bird over almost all of Britain and Ireland, save the highest ground, and had apparently colonised the Outer Hebrides and Orkney during the 19th century. The birds have undoubtedly increased with the spread of human habitation, forestry planting and hedgerows. There is some evidence, from the two Breeding Atlases, that there may have been losses from the north-western mainland of Scotland. The CBC records for the years 1972-96 show declines of 40% on farmland and 41% on woodland. These are serious and unexplained but may be complicated by the very complex social systems with ordinary pairing and almost any combination of up to three or four males and females breeding together! Severe treatment of hedges in farmland has been blamed for local declines. *Still a common bird but the unexplained declines are worrying.*

ROBIN (European Robin)

Erithacus rubecula

> *Distribution* Britain 2,629 (+1.0%) Ireland 967 (-0.9%)
> *Numbers breeding*: Britain 4,200,000 Ireland 1,900,000
> *European status*: 37,000,000 (17% in Britain and Ireland =1)
> *British population trend*: stable or slight increase (+21% CBC){+42%}
> *How likely are you to record it?* 3744 squares (83.4%) *Ranked 6* [3]

Britain's national bird is pretty well ubiquitous, missing only from the remotest islands, the highest mountain and the most open habitats without trees or shrubs. There seems to have been little, if any, change in distribution save a gradual increase in Orkney and on the Western Isles as cover grows up. They do not regularly breed on Shetland. The effects of cold winters are felt on the population levels for a year or two but the birds are better at avoiding losses than many species – a 50% decline seems to be the maximum. Marginal losses have been suffered in arable areas as hedges were lost but new housing must have created so many gardens that these losses will not have had an overall effect. There has been a recent increase in the CBC index and the positive change of 5% in the BBS is significant too. *A species whose prospects seem to be 'set fair'.*

NIGHTINGALE (Common Nightingale)

Luscinia megarhynchos

> *Distribution* Britain 457 (-28.5%) Ireland 0
> *Numbers breeding*: Britain 5,500 Ireland 0
> *European status*: 3,700,000 (0% in Britain and Ireland)
> *British population trend*: declining (-42% CES adults)
> *How likely are you to record it?* 65 squares (1.4%) *Ranked* 108=

Although there is one record of a bird in song in Ireland and a handful in Scotland, Nightingales have not bred in either country. A hundred years ago the birds bred mainly south and east of the line from the Wash to the Severn – none in Cornwall and only a few in Southern Devon. They were regular in small numbers in eastern Wales and north to Cheshire, Nottinghamshire and East Yorkshire. Numbers (and range) seemed to gradually decline until about 1930 accelerating later. The last birds to breed in Wales were recorded in 1981. By the second Breeding Atlas there had been a net loss of 28.5% of 10-km squares. The survey completed in 1999 recorded 4,407 singing males. The estimate for the second Breeding Atlas was 5,000 to 6,000 with surveys of males resulting in 3,230 found in 1976 and 4,770 in 1980. The strongholds are in Sussex, Kent, Essex, Suffolk and Norfolk. The CES results, with significant declines in both adults (-42%) and young (-62%) birds caught, are very serious. The losses have probably several causes. In Britain the birds are at the limit of their range. Coppice and woodland management sympathetic to the birds has gradually diminished. The spread of deer has destroyed the thick ground cover needed where they prefer to breed. *A very worrying loss from many areas but they seem to be relatively safe in some.*

BLUETHROAT

Luscinia svecica

Singing birds have several times been recorded, particularly in Scotland, and a nest with eggs and female were recorded in the Spey Valley in 1968 (no male seen). The RBBP also have a successful record for 1995. This was of the race *svecica* but colonisation of East Anglia or Kent by the white-spotted form *cyanecula* from the expanding population in the Netherlands is also possible.

BLACK REDSTART

Phoenicurus ochruros

> *Distribution* Britain 103 (+51.5%) Ireland 0
> *Numbers breeding*: Britain 99 RBBP Ireland 0
> *European status*: 4,500,000 (0% in Britain and Ireland)
> *British population trend*: gradually increasing
> *How likely are you to record it?* 4 squares (0.1%) *Ranked 167=*

This was a very rare visitor in the 19th century but was recorded nesting in Durham in 1845 and Sussex in 1909 – and two pairs close by in 1923 on coastal cliffs. This sort of site continues to be used but in 1926 the first city site was used inland – extensive destruction by the Luftwaffe(in World War II) provided extra breeding sites. The number of breeding pairs fluctuated from about 20 to 70 pairs up to 1970 and increased to over 100 in some years recently (118 in 1988 for instance). There are now fewer in London and along the South Coast and more in the Midlands and East Anglia. Urban sites used are now most likely power stations, marshalling yards, warehousing estates, etc., and not bombed sites. Breeding has been suspected on about five occasions in Scotland since 1973 and may, possibly, have taken place in Ireland as there have been a few summer records. The first breeding in South Wales was in 1981 and in the North 1984. *Now properly established as a British breeding bird.*

REDSTART (Common Redstart)

Phoenicurus phoenicurus

> *Distribution* Britain 1,327 (-20.1%) Ireland 11 (+22.2%)
> *Numbers breeding*: Britain 90,000 Ireland <5
> *European status*: 2,300,000 (4% in Britain and Ireland =6)
> *British population trend*: good recent recovery (+109% CBC){+30%}
> *How likely are you to record it?* 307 squares (6.8%) *Ranked 72*

The Redstart numbers breeding in Britain seem to have a history of periodic fluctuations – for instance 1891 was a very good year in Scotland but the following year was very bad. There was undoubtedly an overall increase and extension in Northern Scotland at that time but they dwindled in lowland Britain – Wales and near counties of England were the stronghold. The birds were absent from the islands (no forests) and have always been rare in Ireland: most are recorded in County Wicklow.

There is a continuing withdrawal from lowland England. In some areas (like the Vale of Aylesbury) they used to nest in mixed farmland with old trees in large hedges but the area was deserted about 30 years ago. At this time the species was in severe decline for about 10 years but the recovery, over the next 10 years, seems to have been excellent. Recent population and NRS statistics are good. *Fluctuations likely but prospects good.*

WHINCHAT

Saxicola rubetra

Distribution Britain 1,404 (-16.2%) Ireland 124 (-35.1%)
Numbers breeding: Britain 21,000 Ireland 1,900
European status: 2,700,000 (1% in Britain and Ireland)
British population trend: gradually declining
How likely are you to record it? 183 squares (4.1%) *Ranked* 91= [88=]

Whinchats were very common birds over most of mainland Britain 100 years ago breeding in rough pasture, waste ground, roadside verges, railway embankments, etc. They were not particularly common in Cornwall, few breeding in the Western Isles and none on Shetland or the Isle of Man. In Ireland most were in the northern half. A decline in lowland Britain started sixty or seventy years ago and this has continued with many losses between the two Breeding Atlases so that now the strongholds are in the uplands of Wales, northern England and throughout Scotland. Breeding is now regular in the Isle of Man and on the Western Isles but not on Orkney. In Ireland the birds seem to have shifted to the Midlands by 1950 and then increased generally by the first Breeding Atlas, but declined badly by the second. Many of the losses in lowland areas are probably associated with early cutting of the grass, where they breed, and general tidying up. *Further losses from lowland Britain and Ireland seem to be expected.*

KB.

STONECHAT

Saxicola torquata

> *Distribution* Britain 1,034 (-14.6%) Ireland 569 (-28.2%)
> *Numbers breeding*: Britain 15,000 Ireland 13,000
> *European status*: 1,500,000 (2% in Britain and Ireland =10)
> *British population trend*: overall declines
> *How likely are you to record it?* 133 squares (3.0%) *Ranked 96= [39]*

The Stonechat was common more or less throughout Britain and Ireland a century ago with slight declines recorded from a few areas, mainly in the Home Counties. However these losses quickly accelerated, but increases were reported from north and west Scotland. The loss of heathland and other wastes was the clear reason. Stonechats do not move very far in winter and their populations may be hit very hard by winter cold. Where their breeding habitat becomes dispersed the birds are likely to diminish further and this has clearly happened. Now the coastal fringe of East Anglia and South Coast, New Forest and Surrey Heaths are the only population centres in central and eastern Britain. There have also been widespread losses inland in Eastern Ireland and elsewhere. These might indicate that more are going to be lost. *Even after a succession of mild winters Stonechats have not returned to areas like the Brecks – we may be about to lose more!*

WHEATEAR (Northern Wheatear)

Oenanthe oenanthe

> *Distribution* Britain 1,738 (-6.8%) Ireland 433 (-19.8%)
> *Numbers breeding*: Britain 55,000 Ireland 12,000
> *European status*: 3,000,000 (2% in Britain and Ireland = 8=)
> *British population trend*: seems stable after decline
> *How likely are you to record it?* 514 squares (11.4%) *Ranked 62 [48=]*

A hundred years ago Wheatears bred all over the country using roadside pits and quarries and rabbit burrows if other nest sites were not available. In some areas huge numbers had been caught for the table but this had stopped by then. Not much change was recorded until 1940 when declines started to be noticed in peripheral areas and, particularly, in the southern half of England. This has accelerated and few areas south and east of the Humber to Severn line now have breeding Wheatears. Away from the coast and upland areas there have also been losses between the Breeding Atlases in Ireland. The 45% increase in BBS index over five years is encouraging. In many areas the habitats suitable for breeding Wheatears are not now present. *The reduced area with breeding Wheatears is probably as good as it will get.*

RING OUZEL

Turdus torquatus

> *Distribution* Britain 544 (-27.0%) Ireland 29 (-17.1%)
> *Numbers breeding*: Britain 8,000 Ireland 270
> *European status*: 280,000 (3% in Britain and Ireland =8)
> *British population trend*: continuing serious declines
> *How likely are you to record it?* 45 squares (1.0%) *Ranked* 118=

Ring Ouzels were widely distributed through upland areas of Britain and much of Ireland, but not Shetland or the Outer Hebrides, a hundred years ago. However a dramatic and progressive decline set in and the birds retreated to the remotest uplands. The retreat has been particularly marked in Ireland where there are now only three areas with more than four occupied 10-km squares in the latest Breeding Atlas (seven other records). In Scotland the birds have also been lost from many areas – including Orkney and Arran. In Wales the breeding population may number 300 to 400. The declines are blamed on forestry, disturbance by walkers, agricultural improvement and the Ring Ouzel losing out to Blackbird and Mistle Thrush established before the migrants return in spring. The net loss of 27% of occupied 10-km squares between the two Breeding Atlases is very worrying. *Further losses seem inevitable.*

BLACKBIRD (Common Blackbird)

Turdus merula

> *Distribution* Britain 2,664 (-1.9%) Ireland 976 (-0.8%)
> *Numbers breeding*: Britain 4,400,000 Ireland 1,800,000
> *European status*: 43,000,000 (14% in Britain and Ireland =2)
> *British population trend:* decline possibly stabilising? (-33% CBC){-29%}
> *How likely are you to record it?* 3949 squares (87.9%) *Ranked* 3 [2]

This very common bird spread from its woodland and woodland edge habitat to breed in gardens in towns and cities and extensively through farmland. By the end of the 18th century it had also expanded into the northernmost parts of Scotland, to the Western Isles and Orkney – and throughout Ireland. It may seem incredible that this is a bird causing concern because of reduction in numbers. However the data from the

CBC shows a quick recovery from the effects of the cold winter of 1962/63 and then rather steady populations for 15 years. However the 25-year change from all CBC sites is a reduction of 33% (1972-96) – easily enough to set the alarm bells ringing. A careful and detailed analysis of the CBC data for individual plots can tease no causative relationship between loss of Blackbirds and the increase in Magpies and Sparrowhawks. Detailed investigation of survival rates of ringed birds shows that these are low when the population is declining but much more work needs to be done to find out how this decline is driven. *Perhaps the decline is over, since the BBS does not show a significant change, but do not count on it!*

FIELDFARE

                                   Turdus pilaris

> *Distribution* Britain 104 (+205.9%) Ireland 0
> *Numbers breeding*: Britain 2 RBBP Ireland 0
> *European status*: 6,000,000 (0% in Britain and Ireland)
> *British population trend:* just a toehold at the moment
> *How likely are you to record it?* 65 squares (1.4%) *Ranked* 108 [101=]

This big thrush has steadily increased westwards from its normal range and first bred in Orkney during 1967 (and probably Durham the same year). Small scale Scottish breeding, particularly on Shetland, became regular. Later the birds were found on the mainland south to the South Pennines and 10 or more pairs were detected in each year from 1989 and 1992. First recorded breeding in Kent in 1991 indicated Central European birds might be coming but only two pairs found in 1997. *Still on their way?*

SONG THRUSH

Turdus philomelos

> *Distribution* Britain 2,620 (-2.1%) Ireland 947 (-2.3%)
> *Numbers breeding*: Britain 990,000 Ireland 390,000
> *European status*: 16,000,000 (9% in Britain and Ireland = 3=)
> *British population trend:* serious losses (-52% CBC, -66% farm){-55%}
> *How likely are you to record it?* 2,944 squares (65.6%) *Ranked* 14 [9]

The Song Thrush was a familiar and common bird throughout Britain and Ireland where there was some sort of cover a hundred years ago. It was absent from Shetland and sparse in the far west of Ireland and, every now

and then, knocked back in numbers by cold winter weather. These losses were generally made good after a few years of mild winters. Earlier this century they colonised Shetland but, since the very cold winter of 1947, they are not regular there. There is some evidence, from 60 or 70 years ago, that the ratio of Song Thrush to Blackbird began to change very much in the Blackbird's favour. However it was 25 years ago that the CBC index began to show a steep decline (-66% on farmland) and, although they are still common birds, many fewer are present particularly in farmland and garden areas. In some areas the number of breeding attempts each year is now too few to sustain the population. Overall it seems that the survival of young birds through their first winter has declined in the last 20 years. These subtle changes may be enough to cause the declines. *Is the 1998 increase of 18% on the BBS going to be sustained? It will need to carry on for several years to make good the losses!*
UKBAP **English Nature** RSPB.

REDWING

Turdus iliacus

> *Distribution* Britain 136 (+22.5%) Ireland 4 (0%)
> *Numbers breeding*: Britain 17 RBBP Ireland 0
> *European status*: 5,700,000 (0% in Britain and Ireland)
> *British population trend:* relatively stable
> *How likely are you to record it?* 11 squares (0.2%) *Ranked* 148= [101=]

These birds were first recorded breeding in Sutherland in 1925 and sporadically for the next 40 years (and apparently in Kerry in 1951). Fieldwork for the first Breeding Atlas discovered many more and several dozen possible breeding birds were recorded until 1989. However the average has dropped by about half recently even though breeding has been proved in England (particularly Kent). It is possible that this species is being under-recorded. *Almost certainly here to stay.*

MISTLE THRUSH

Turdus viscivorus

Distribution Britain 2,397 (-2.0%) Ireland 857 (-8.8%)
Numbers breeding: Britain 230,000 Ireland 90,000
European status: 2,500,000 (13% in Britain and Ireland =3)
British population trend: declining (-34% CBC, -48% farm){-21%}
How likely are you to record it? 2042 squares (45.5%) *Ranked* 28 [22]

The Mistle Thrush spread North through mainland Scotland and into the Inner Hebrides, and through Ireland, during the 19th century. By 100 years ago the distribution was little different from the modern maps – none breed in Shetland and there are a few, sporadic, records from Orkney and the Western Isles. The birds are badly affected by cold winters but recovery, even after a 75% loss after 1962/63, was fairly quick. However in the 25 years from 1972 the CBC index (all plots) fell by more than a third – and over half on farmland. Current expectation is that this may be due to poor survival over winter. This is another species where much more detailed work is needed and such factors as the number of breeding attempts each year needs to be investigated. *The species seems to be on a downwards pointing roller coaster.*

CETTI'S WARBLER

Cettia cetti

Distribution Britain 86 (+1620%) Ireland 0
Numbers breeding: Britain 574 Ireland 0
European status: 700,000 (0% in Britain and Ireland)
British population trend: new and still increasing
How likely are you to record it? 9 squares (0.2%) *Ranked* 151=

Unknown in Britain until 1961 this skulking (but *very loud*) resident warbler had rapidly spread north from its native Mediterranean sites. Breeding was established in Kent in 1972 and on Jersey in 1973. The colonisation of southern England continued apace but the population centre shifted to Hampshire and Dorset when cold winters virtually wiped out the eastern birds in the 1980s. South Wales was colonised in about 1980 and there was a singing bird in Anglesey in 1997. Kent has been re-colonised and the RBBP maximum has been over 300 for 1993-1997. The maximum was 593 (including Jersey) during a full census in 1996 when the top counties were Hampshire (maximum 143 singing males),

Dorset (97), Devon (88) and Somerset (67). The species will still be vulnerable to very cold winters. *These birds established themselves very quickly and look to be secure.*

GRASSHOPPER WARBLER

 *Locustella naevia*

> *Distribution* Britain 1,189 (-36.7%) Ireland 409 (-40.9%)
> *Numbers breeding*: Britain 10,500 Ireland 5,500
> *European status*: 330,000 (5% in Britain and Ireland = 6=)
> *British population trend*: serious widespread declines (-59% CBC){-73%}
> *How likely are you to record it?* 164 squares (3.7%) *Ranked* 94 [53]

One hundred years ago the song of this monotonous warbler, much more often heard than seen, was heard through mainland Britain as far north as the Highland fringe and over most of Ireland except the north-west coast. Over the next 50 years the birds spread through the rest of mainland Britain, parts of the Inner Hebrides, onto the Isle of Man and increased throughout most of Ireland. The plantation breeding sites are useful for about ten years after planting but the traditional sites of damp tangled vegetation are also likely to be drained. This species apparently suffered huge losses between the two Breeding Atlases. Despite a peak in 1980 the general CBC dropped by 59% between 1972 and 1996. Habitat loss through drainage and maturing forestry has been blamed but the birds may be in trouble in their wintering area which seems to be close to the southern edge of the Sahara. The one good omen is a doubling of registrations on the BBS between 1994 and 1998! *The BBS increase may herald a welcome return but it is rather too early to be certain.*

RIVER WARBLER *Locustella fluviatilis*

Eight records of singing males, a few staying for more than a day or two, in the last 25 years but never suspected of breeding.

SAVI'S WARBLER

 Locustella luscinioides

> *Distribution* Britain 27 (+125%) Ireland 2 (0%)
> *Numbers breeding*: Britain 5 RBBP Ireland 0
> *European status*: 180,000 (0% in Britain and Ireland)
> *British population trend:* – about to be lost?

The Savi's Warbler was not realised to be a distinct species until Savi described it in 1824 (in Italy). It was first recorded, breeding in the Cambridgeshire fens in 1840, but within 20 years it was extinct. A singing bird was recorded in 1854 and breeding commenced in Kent in 1960 and Suffolk in 1970. The population may have reached more than 20 pairs with a couple of records of singing males in Ireland. However for the three years 1995 to 1997 the maxima have been 3, 3 and 5. *Sadly this species may be on the way out as the Dutch population has been declining too.*

MOUSTACHED WARBLER

Acrocephalus malanopogon

Bred in Cambridgeshire during 1946! Otherwise recorded only in 1951, 1952 and 1965.

SEDGE WARBLER

Acrocephalus schoenobaenus

Distribution Britain 1,887 (-8.7%) Ireland 681 (-20.5%)
Numbers breeding: Britain 250,000 – Ireland 110,000
European status: 2,300,000 (15% in Britain and Ireland =3)
British population trend: decline now stabilised (-12% CBC){-45%}
How likely are you to record it? 530 squares (11.8%) *Ranked* 61 [36]

The general distribution of Sedge Warblers in Britain and Ireland has not changed much although they may have colonised more of the Hebrides in the last 50 or 60 years. First breeding on Shetland was proved in 1996 However there are very clear indications of general losses from much of the range between the two Breeding Atlases – possibly more from the west and north of England and from the southern half of Ireland than other areas. The populations seemed to be at a high level in the early 1960s but they suffered a substantial decline in 1969 after the winter drought in their sub-Saharan wintering area. The current 25-year CBC starts after this decline and shows no sign of a proper recovery. Detailed research shows a clear relationship between the survival rates of adult Sedge Warblers and the rainfall index for the wintering area and, unless this consistently improves, a recovery to previous populations seems unlikely. *No real sign of increase to previous levels at the moment but, thank goodness, no sign of further steep declines.*

MARSH WARBLER

 *Acrocephalus   palustris*

Distribution Britain 15 (-28.6%) Ireland 0
Numbers breeding: Britain 32 RBBP Ireland 0
European status: 1,700,000 (0% in Britain and Ireland)
British population trend: possibly increasing
How likely are you to record it? 3 squares (0.1%) *Ranked* 171=

A hundred years ago it was recognised that this species mainly bred in the Severn Valley with a few sporadically breeding in adjacent counties and Hampshire. Later there may have been an increase as birds were found in the south-east corner of England as well. However by 1950 the birds had started to decline although they were still concentrated along the Severn. The first Breeding Atlas recorded 50 – 80 pairs and considered 75% were in Worcestershire with very few in the South-east. Now the Severn population is very weak and the majority are in Kent and Sussex and the numbers are often down to a maximum of less than 30 (but 58 in 1993). Marsh Warblers do breed sporadically elsewhere (including Wales) and there are even two records on Orkney (1993 and 1997). The population is very strong in the Low Countries. *The species may be lost from its traditional areas but the new ones look very good.*
Meek, E.R & Adam, R.G. 1997 *British Birds*: 90, 230.
UKBAP. **Environment Agency** – RSPB & Wildlife Trusts.

REED WARBLER (Eurasian Reed Warbler)

 Acrocephalus   scirpaceus

Distribution Britain 790 (+1.9%) Ireland 13
Numbers breeding: Britain 60,000 + Ireland ~50
European status: 3,100,000 (2% in Britain and Ireland =7)
British population trend: excellent increases (+228% CBC){+194%}
How likely are you to record it? 193 squares (4.3%) *Ranked* 88

By the end of the 19th century the Reed Warbler had spread from breeding solely in the southern third of England northwards to include almost all Lancashire and much of Yorkshire. Later the birds retreated with only

isolated colonies reported from north of the line connecting the Humber and Mersey and sparsely in eastern Wales and Devon. The situation changed little until the first Breeding Atlas when they were a little further north, in Anglesey, along the south coast of Wales and a few places in Devon and Cornwall. Then a real spread started although breeding on Shetland in 1973 may have been from Scandinavia. Extra sites in northern England were colonised and also southern Scotland, more sites in West Wales and the South-west are now regularly used. However the real change is in Ireland where one pair bred in 1953 (Co. Down) but over the last twenty years the breeding population has built up to about 100 pairs. Regular sites are in Cork, Wexford and Wicklow and there are many more reed bed sites for them to find. In England the proliferation of gravel pits and small nature reserves with reeds has clearly provided the Reed Warbler with many extra, and protected, breeding sites – CBC up 228% indicates this. *Further expansion and increase very likely – particularly if the summers get better as this is a late migrant.*
Bruce, K. 1997 *Scottish Birds*: **19**, 119-120.
Smiddy, P. & O'Mahony, B. 1997 *Irish Birds*: **6**, 23-28.

GREAT REED WARBLER *Acrocephalus arundinaceus*

Singing males are now regularly reported from reed beds (even one in Scotland) and twice at the same site in successive years in Kent, the birds have been seen carrying nesting material – nonetheless no record of breeding. *Yet!*

ICTERINE WARBLER

Hippolais icterina

There are a couple of records about 100 years ago that may be of this species breeding in England. In 1970 they may have bred in Yorkshire and were proved in 1992 on the mainland of Northern Scotland.

DARTFORD WARBLER

Sylvia undata

> *Distribution* Britain 45 (+60.7%) Ireland 0
> *Numbers breeding*: Britain 1,679 Ireland 0
> *European status*: 2,600,000 (0% in Britain and Ireland =5)
> *British population trend*: seriously increasing and expanding
> *How likely are you to record it?* 8 squares (0.2%) *Ranked* 155=

The Dartford Warbler has always been associated with gorse scrub and as a largely resident small insectivorous bird was subject to huge losses after cold winter weather. At the end of the 19th century, before huge losses following two cold winters, they were present in all counties south of the Thames and west to Dorset with a very few in Cornwall, Shropshire and coastal Suffolk. Later they were largely confined to Sussex, Surrey and Hampshire. Fluctuating populations due to cold winters and the loss of heathland habitat led to a maximum of about 450 pairs immediately before the two cold winters of 1961/62 and 1962/63. These knocked the population back to ten pairs in 1963 (excluding the Channel Isles) and these had more than doubled, to 22, by 1966. After a good series of mild winters over 900 were estimated by 1990 – only a couple of dozen outside of Hampshire, Dorset and Surrey. A national survey found 1600 pairs in 1994 and there may have been more by 1997 when, despite the loss of many areas of heathland, territories have been found in Avon, Berkshire, Norfolk and Suffolk as well as the core areas. *Soon Kent (one pair 1995) will have this warbler named after one of their heaths back and regularly breeding – provided the winters stay mild!*

SUBALPINE WARBLER *Sylvia cantillans*

A few singing males reported recently have seldom stayed more than a day or two and are clearly not really trying!

LESSER WHITETHROAT

Sylvia curruca

> *Distribution* Britain 1,271 (+16.2%) Ireland 1
> *Numbers breeding*: Britain 80,000 Ireland ~1
> *European status*: 2,200,000 (4% in Britain and Ireland =9)
> *British population trend*: stable after decline (-24% farm CBC){+17%}
> *How likely are you to record it?* 365 squares (8.1%) *Ranked 67*

The Lesser Whitethroat was much confused with the Whitethroat earlier but by the end of the 19th century its distribution was established as very much the same as for the first Breeding Atlas. The birds migrate to the South-east and were virtually absent from South Wales, Devon and Cornwall and just crept into Scotland (possibly absent for some time around 1960s) and West Wales. The birds expanded much more by the second Atlas. Several areas in Scotland have established populations, breeding is more extensive in northern England and West Wales and singing birds have been found in Orkney, Shetland, the Isle of Man and even Ireland: where there have been recent winter records. However a very stable CBC graph showed a small scale decline in the late 1980s and the 5-year BBS index is down significantly by 36%. *Are there troubles in the wintering area? A species to watch and worry about.*

WHITETHROAT (Common Whitethroat)

Sylvia communis

> *Distribution* Britain 2,186 (-6.7%) Ireland 628 (-25.5%)
> *Numbers breeding*: Britain 660,000 + Ireland 120,000
> *European status*: 7,300,000 (11% in Britain and Ireland =2)
> *British population trend:* recovering following slumps (+83% CBC){+34%}
> *How likely are you to record it?* 2176 squares (48.5%) *Ranked* 24 [45]

A very widespread migrant, the Whitethroat had apparently spread North in Scotland and, at the end of the 19th century was present throughout the mainland (save the very north), most of the Inner and parts of the Outer Hebrides. The expansion continued with increases and breeding, for a few years, on Orkney. The birds were well distributed throughout Ireland wherever there was suitable cover. The summer of 1969 saw all this change with massive winter mortality attributed to the Sahel drought. The CBC index dropped by 75% over winter and there was a further 50% decline over the next five years. The gradual recovery was reversed (same reason) in 1984-85 but subsequently prospects have seemed good and the birds have even bred on Shetland. The 5-year BBS figures show a significant increase of 14%. *Prospects good BUT still vulnerable to droughts in the Sahel.*

GARDEN WARBLER

Sylvia borin

> *Distribution* Britain 1,867 (+1.9%) Ireland 56 (+5.7%)
> *Numbers breeding*: Britain 200,000 Ireland 240
> *European status*: 11,000,000 (2% in Britain and Ireland)
> *British population trend:* increasing (+27% CBC){+35%}
> *How likely are you to record it?* 848 squares (18.9%) *Ranked* 53 [84=]

By the end of the 18th century it had been established that the Garden Warbler was breeding over most of England and Wales and into Scotland south of the Highlands. It was absent from the Isle of Man but bred in several small pockets of suitable habitat over much of Ireland. The situation remained similar for 50 or 60 years and then breeding started to be recorded in the Isle of Man and further north in Scotland. By the time of the second Breeding Atlas there were further gains in Scotland but birds were lost from almost 30 10-km squares in Devon and Cornwall. Fluctuations in the CBC index include a major gradual loss in the 1970s

and a steady recovery up to about 1990. These changes may have been driven by problems on the wintering grounds. *Prospects seem good provided the winter weather in Africa holds out.*
Lovatt, J. 1997 *Irish Birds*: 6, 58-60.

BLACKCAP

Sylvia atricapilla

> *Distribution* Britain 2,048 (+6.5%) Ireland 357 (+39.5%)
> *Numbers breeding*: Britain 580,000 + Ireland 40,000
> *European status*: 21,000,000 (3% in Britain and Ireland)
> *British population trend:* long-term increase (+54% CBC){+155%}
> *How likely are you to record it?* 2291 squares (51.0%) *Ranked* 20 [50=]

A hundred years ago the Blackcap was very common in the Southern two thirds of England and over most of Wales except the far West. There were fewer birds in the rest of England and north to the Highland fringe and, in Ireland, there were scattered groups of birds in several areas. They may have withdrawn from many areas in Ireland earlier this century but by 1970 they were found to breed widely over the eastern half of the country and there was further consolidation and expansion west. In Scotland the birds spread North and now breed down the Great Glen and have even been recorded nesting a few times on Orkney and Shetland. British Blackcaps generally winter in the Mediterranean and escape the ravages of local cold winters and the effects of Sahel droughts. The CBC shows consistent increases which are clearly continuing as the 1994-98 BBS increased by 42%! *A species on the up and up.*

WOOD WARBLER

Phylloscopus sibilatrix

> *Distribution* Britain 1,270 (+3.3%) Ireland 28 (+250%)
> *Numbers breeding*: Britain 17,200 — Ireland 30
> *European status*: 6,500,000 (0% in Britain and Ireland)
> *British population trend:* some losses recently (-43% BBS)
> *How likely are you to record it?* 133 squares (3.0%) *Ranked* 96

Wood Warblers were known to breed in appropriate woodland throughout the mainland of Britain at the end of the 19th century and sporadically sang in just a few places in Ireland. In the previous 100 years there had been a considerable spread North in Scotland and this continued gradually until about 1950 since when there has probably

been a gradual withdrawal. There were few breeding records in Ireland until 1968 when breeding was proved in Wicklow and it gradually became more regular there and from 1980 in up to half a dozen or so other areas. The second Breeding Atlas recorded many more 10-km squares with records in Ireland and some more in South-west Wales and through Scotland but many losses from South-east Britain. There are now very few North of the Thames and east of a line from the Severn to the Humber; possibly a third of Britain's birds breed in Wales. CBC indices show fluctuations but there is recent evidence for a very worrying decline of 43% in the five-year BBS index (1994-1998). *On its way out from East Anglia and some other areas.*

CHIFFCHAFF (Common Chiffchaff)

Phylloscopus collybita

Distribution Britain 2,100 (+5.0%) Ireland 836 (-7.3%)
Numbers breeding: Britain 640,000 + Ireland 290,000
European status: 16,500,000 (6% in Britain and Ireland = 7=)
British population trend: recently increasing (+32% BBS){+26%}
How likely are you to record it? 2160 squares (48.1%) *Ranked* 27 [26]

There was a big expansion of breeding Chiffchaffs during the 19th century through England and into Scotland as well as throughout Ireland and Wales (few in Anglesey). By 100 years ago they were just about everywhere, in suitable habitats, north to the Lowlands of Scotland. Expansion continued through Ireland and onto some of the Irish islands and through the Scottish mainland and into the Inner Hebrides. New areas were colonised between the two Breeding Atlases including singing birds on the Western Isles and Orkney but they were lost from some areas of Western Ireland. The CBC showed a large increase following the cold winter of 1962/63 although most winter in the Mediterranean. However there was a subsequent loss in the early 1970s made up in the late 1980s so that the highest levels were regained – and there has been a 32% increase in the five-year BBS index. *Probably at their highest population level at the moment.*

WILLOW WARBLER

Phylloscopus trochilus

Distribution Britain 2,602 (+0.6%) Ireland 927 (-1.4%)
Numbers breeding: Britain 2,300,000 Ireland 830,000
European status: 39,000,000 (8% in Britain and Ireland = 4=)
British population trend: stable until recent declines (-23% CBC){-23%}
How likely are you to record it? 2706 squares (60.3%) *Ranked* 16 [17]

The Willow Warbler was considered, 100 years ago, to be the commonest warbler in northern England and over most of Scotland but the Whitethroat was more numerous in England. They were sparse in the west of Ireland and breeding was not proved on Orkney or the Western Isles and not suspected on Shetland. Now the birds have expanded to occupy extra areas in western Ireland and to breed on Orkney and the Western Isles – even, sometimes, on Shetland. Everything seemed to be going very well with a very steady CBC index until about ten years ago when there were bad losses which affected the southern British population but not those in the North. Detailed research indicates that these were due to poor adult survival and thus changes on migration or in Africa might have been to blame. More recently the five-year BBS index (1994-1998) logged a strong recovery of 25% so the problem may only be a short-term glitch. This is a species showing a significantly earlier first egg date and an increasing brood size – probably correlated with global warming. *Looking quite good but are there clouds on the horizon still?*

GOLDCREST

 Regulus regulus

> *Distribution* Britain 2,327 (-0.6%) Ireland 852 (-6.6%)
> *Numbers breeding*: Britain 560,000 Ireland 300,000
> *European status*: 10,500,000 (8% in Britain and Ireland =4)
> *British population trend*: declining badly (-60% CBC){0%}
> *How likely are you to record it?* 1099 squares (24.5%) *Ranked* 43 [23=]

By the end of the 19th century the Goldcrest seems to have been breeding virtually throughout mainland Britain and Ireland – and also on the Inner Hebrides – following expansion due to new forestry plantations maturing in many areas. Expansion in the Western Isles from breeding in Stornoway Woods, on Orkney from 1945 and even on Shetland has been recorded. The populations have always been very vulnerable to cold winter weather and the CBC index seems to have increased ten-fold after 1962/63. However this peak has never been regained and the smoothed CBC index has been falling steadily over the last 20 years. There have been one or two rather cold winters during this period but the birds do not seem to have responded to the many mild winters that intervened in the way expected. The only good news is that the five-year BBS rose by 42% between 1994 and 1998. The density map in the second Breeding Atlas shows that there are few in England North of the Thames compared with other areas. *Perhaps we should be getting seriously worried about the lowland populations of these birds.*

FIRECREST

 Regulus ignicapillus

> *Distribution* Britain 99 (+395%) Ireland 0
> *Numbers breeding*: Britain 60 Ireland 0
> *European status*: 3,800,000 (0% in Britain and Ireland)
> *British population trend*: fluctuating – but established
> *How likely are you to record it?* 1 square (0.0%) *Ranked* 190=

These tiny birds were first satisfactorily proved to breed in Britain in the New Forest in 1962 and they were found in Hertfordshire, Kent and Dorset by 1970. It became clear that the colonisation had spread much wider with breeding recorded at many sites from South Wales to the Wash and southwards as well as other parts of Wales and north to Lancashire and Derbyshire. There are no signs of breeding in Scotland or Ireland. Numbers are very different from year to year with a maximum of 131 in 1989 and a minimum of 19 in 1992 (after a cold winter) being the extremes for 1988 to 1997. *This seems now to be an established British breeding bird.*

SPOTTED FLYCATCHER

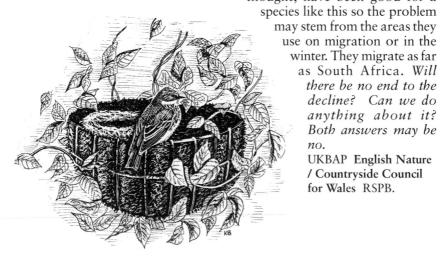 *Muscicapa striata*

> *Distribution* Britain 2,378 (-2.3%) Ireland 728 (-18.2%)
> *Numbers breeding*: Britain 120,000 – Ireland 35,000
> *European status*: 7,800,000 (2% in Britain and Ireland)
> *British population trend:* declining very seriously (-78% CBC){-68%}
> *How likely are you to record it?* 441 squares (9.8%) *Ranked* 64 [57=]

The Spotted Flycatcher needs a good supply of flying insects and perches to sally forth and catch them, so it is absent from really open countryside. At the end of the 19th century they were breeding over the entire mainland of Britain, and the Inner Hebrides, and Ireland. About 50 years ago they started to breed on the Western Isles and now they also nest sporadically in Orkney. Between the two Breeding Atlases there were losses – especially from the west of Ireland. There were reports of declines in the 1950s later confirmed by the CBC. Initially down by about a third they have declined by 78% over the 25 years 1972-96 and the BBS (1994-98) shows a significant drop of 23%. Global warming would, one would have thought, have been good for a species like this so the problem may stem from the areas they use on migration or in the winter. They migrate as far as South Africa. *Will there be no end to the decline? Can we do anything about it? Both answers may be no.* UKBAP **English Nature / Countryside Council for Wales RSPB.**

PIED FLYCATCHER

Ficedula   hypoleuca

> *Distribution* Britain 732 (+34.1%) Ireland 3 (0%)
> *Numbers breeding*: Britain 37,500 Ireland <5
> *European status*: 5,250,000 (1% in Britain and Ireland =9)
> *British population trend*: recent evidence of declines
> *How likely are you to record it?* 98 squares (2.2%) *Ranked* 105

The Pied Flycatcher can be the dominant breeding species in Welsh oak woodlands now that a super-abundance of nest boxes are present in many areas but it was not always so. A hundred years ago North Wales and North-west England were the undoubted strongholds with birds also breeding in mid-Wales, the Welsh Marches, other parts of northern England and on the Scottish mainland (predominately eastern areas). In the last 30 years they have gained a foothold in Ireland with breeding reported mainly in Armagh and Wicklow. In Scotland the main areas are in Dumfries and Galloway and in the Trossachs but they bred sporadically in many other areas. About 50 years ago it was clear that the birds had become well established in mid-Wales and were starting to breed in South-west England. The second Breeding Atlas documented further gains in South-west Wales and England, along the eastern edge of the Welsh breeding area and in northern England. Recently there have been several reports of generally lower numbers using long-established nest box schemes. Fluctuating numbers, possibly because of winter losses or poor productivity, would be nothing new but poor results for over ten years seem to be unprecedented. *These reports give cause for serious concern.*

BEARDED TIT

Panurus   biarmicus

> *Distribution* Britain 60 (+33.3%) Ireland 2
> *Numbers breeding*: Britain 370 Ireland 0
> *European status*: 250,000 (0% in Britain and Ireland)
> *British population trend*: seems relatively stable
> *How likely are you to record it?* 3 squares (0.1%) *Ranked* 171=

A hundred years ago this reed-bed species was restricted to East Anglia – possibly just Norfolk – having previously been in many other counties where drainage had destroyed their habitat. Previous areas included the Severn, the Humber and Hampshire, Surrey, Sussex and Kent. Subsequently the Suffolk reed beds were re-colonised but the succession

of cold winters between 1917 and 1963 hammered them – after 1947 they may have been down to two to four pairs! The 1962/63 winter did not affect the population as badly and immigration from the huge populations breeding on the Dutch polders partly fuelled a considerable expansion to include Humberside, Dorset, Hampshire, Sussex, Kent, Hertfordshire and Essex by the time of the first Breeding Atlas. By the second Atlas they had reached Wicklow (Ireland) for about 10 years and at least two sites in Wales but then retreated from them. However Leighton Moss and many new sites along the South Coast and Thames Estuary were successfully colonised and breeding in Scotland was first recorded in 1991. An estimate of 408 pairs in 1992 is about two-thirds that estimated for 1980 and there may have been a recent increase almost to the level 20 years ago. They are also being helped by ingenious 'wigwams' of reeds which prove to be very successful nest sites! *Vulnerable to cold weather but now breeding in a good number of widely spaced reed beds.*

Campbell, L., Cayford, J. & Pearson, D. 1996 *British Birds*: **89**, 335-346.

LONG-TAILED TIT

Aegithalos caudatus

Distribution Britain 2,106 (-4.0%) Ireland 546 (-24.1%)
Numbers breeding: Britain 210,000 Ireland 40,000
European status: 3,100,000 (8% in Britain and Ireland =4)
British population trend: stable or increasing (+19% CBC){+61%}
How likely are you to record it? 1278 squares (28.5%) *Ranked* 35 [48=]

Breeding Long-tailed Tits were well distributed across Britain and Ireland 100 years ago wherever there was cover for them to nest in. It is possible that they have expanded a little in northern Scotland and the Inner Hebrides since then and there are sporadic records of breeding on the Western Isles. There was not much change in Britain between the two Breeding Atlases but there were losses all over Ireland. This tiny species is very vulnerable to cold winter weather – particularly glazed frosts – and the losses generally take just a few years before complete recovery. The CBC shows a peak in 1974 which has recently been passed. *Provided the winter weather does not deteriorate Long-tailed Tits should do well.*

MARSH TIT

   *Parus palustris*

> *Distribution* Britain 1,133 (-17.1%) Ireland 0
> *Numbers breeding*: Britain 60,000 – Ireland 0
> *European status*: 3,300,000 (2% in Britain and Ireland)
> *British population trend*: declining (-37% CBC){-69%}
> *How likely are you to record it?* 260 squares (5.8%) *Ranked 80=*

A hundred years ago the Marsh and Willow Tits were not separated as two species so the records of 'black-capped tits' refer to both species combined. They were found throughout England and Wales (save Anglesey) and also through Scotland north to the Spey with most records from the South-east and the Lowlands. Neither has ever been recorded in Ireland. Gradually the two species were unravelled and it was realised that the Marsh Tit bred through most of England except Cumberland and there were few records from Scotland. Indeed the first proof of breeding was not until 1945 and there are now 50-100 pairs breeding in the Borders Region. For the second Breeding Atlas there were some gains in Cumbria, Southern Scotland and West Wales but there were net losses. There has been a net loss in the CBC and the only crumb of comfort is a non-significant increase of 15% in the five-year BBS index (1994-98). *Perhaps this species is losing out to the commoner tits?*

WILLOW TIT

   *Parus montanus*

> *Distribution* Britain 1,100 (-9.8%) Ireland 0
> *Numbers breeding*: Britain 25,000 – Ireland 0
> *European status*: 5,000,000 (1% in Britain and Ireland)
> *British population trend*: declining very seriously (-50% CBC){-63%}
> *How likely are you to record it?* 125 squares (2.8%) *Ranked 99*

Once the confusion with Marsh Tits had been sorted out it became clear that the Willow Tit was the species breeding north to the Highland fringe in Scotland and with small populations further north. However it began to retreat southwards several decades ago and has not been recorded regularly breeding north of the Lothians for many years. The second Atlas shows further losses from southern Scotland and all over the rest of its range – especially in Cornwall. The only gain was Anglesey and another major loss was from most of the North-west of England. These very sedentary birds clearly lose out in cold winters but the CBC index shows a 50% decline over 1972 to 1996 and also 30% over the five years 1994-98 in the BBS. Again losing out in competition with the other commoner tit species is a possible cause. *The dull black cap of the Willow Tit recalls the black cap of a judge pronouncing the death sentence – it can't be true? Can it!*

Maxwell, J. 1999 *BTO News*: **221**, 12-13.

CRESTED TIT

Parus cristatus

> *Distribution* Britain 51 (+10.9%) Ireland 0
> *Numbers breeding*: Britain 900 Ireland 0
> *European status*: 4,000,000 (0% in Britain and Ireland)
> *British population trend:* apparently increasing
> *How likely are you to record it?* 3 squares (0.1%) *Ranked* 171=

The Crested Tit probably bred extensively through the Caledonian pine forests of the Highlands of Scotland before they were destroyed but, by the end of the 19th century, they seem to have been confined to a small part of the Spey Valley. At about this time a slow expansion seems to have started with more of the Spey Valley occupied and breeding birds being found in Caledonian forest or Scots pine plantations as far north as Easter Ross and west to Strath Farrar and Glen Affric. *The birds are vulnerable to severe winters but they seem safe and probably set to expand.*

COAL TIT

Parus ater

> *Distribution* Britain 2,315 (-3.2%) Ireland 854 (-0.1%)
> *Numbers breeding*: Britain 610,000 Ireland 270,000
> *European status*: 14,000,000 (6% in Britain and Ireland =5)
> *British population trend:* increasing (+28% CBC, +63% wood){+55%}
> *How likely are you to record it?* 1165 squares (25.9%) *Ranked* 40 [21]

Although appearing in all sorts of woodland the Coal Tit is particularly at home in conifers where its thin bill enables it to forage easily amongst the needles. A hundred years ago, as now, the birds were found everywhere there was woodland and were expanding in areas where conifers had been planted. In Ireland, where the local race *hibernicus* has been separated from the British *britannicus*, it has spread westwards as conifer plantations have been established in otherwise treeless areas. The same has happened in Scotland and there have been a few years when they have bred on the Western Isles (including 1997). Populations are liable to be hit by hard winters but have not been so affected much since 1962/63. The 25-year CBC index (1972-96) is up by 63% and the five-year BBS (1994-98) has increased by 32%. *A species that is doing very well.*

BLUE TIT

Parus caeruleus

Distribution Britain 2,480 (-1.4%) Ireland 930 (-2.0%)
Numbers breeding: Britain 3,300,000 Ireland 1,100,000
European status: 18,000,000 (25% in Britain and Ireland =1)
British population trend: stable or increasing (+28% CBC){+35%}
How likely are you to record it? 3740 squares (83.3%) *Ranked* 7 [10]

Probably overall this has always been the most common tit in Britain and Ireland. At the end of the 19th century it had only just colonised the Isle of Man – probably due to recent increased tree planting – and was still expanding to the very north of the mainland of Scotland. It got to the Western Isles about 35 years ago. In Ireland the second Breeding Atlas showed a few losses along the West Coast but some gains. These birds too are subject to cold weather mortality but the CBC shows a gradual and steady increase from the mid 1960s until 1996 (and BBS). This is despite one or two recent breeding seasons when the timing of breeding has not coincided with the caterpillar flush. This might be a feature of changes to do with global warming. *Unless global warming provides the birds with a problem timing their egg laying, these birds will continue to do well.*

GREAT TIT

Parus major

Distribution Britain 2,443 (-0.5%) Ireland 883 (-2.2%)
Numbers breeding: Britain 1,600,000 Ireland 420,000
European status: 42,000,000 (5% in Britain and Ireland =7)
British population trend: stable or increasing (+22% CBC){+57%}
How likely are you to record it? 3,372 squares (75.1%) *Ranked* 8 [18]

About a hundred years ago Great Tits were breeding throughout wooded areas and those with scattered trees throughout Britain and Ireland except for Scotland North of the Great Glen. In subsequent years they spread northwards and colonised Lewis at the same time as Blues. The second

Breeding Atlas documents further colonisation of Western Mayo and Connemara which may be correlated with new forestry plantings. These birds may not be quite as vulnerable to cold winter weather as their smaller cousins and their CBC index has gradually increased, in general terms, over the last 35 years – and the five-year BBS went up by 14%. As with other garden feeding species they may well have a better potential winter survival now than earlier this century. However, like the Blue Tit, there is some evidence that global warming may cause changes in spring weather that can de-couple their timing of breeding from the woodland caterpillar cycle. *Set fair as a familiar bird throughout the country and in our gardens.*

NUTHATCH (Wood Nuthatch)

Sitta europea

Distribution Britain 1,270 (+8.2%) Ireland 0
Numbers breeding: Britain 130,000 Ireland 0
European status: 7,000,000 (2% in Britain and Ireland)
British population trend: increasing and expanding (+82% CBC){+118%}
How likely are you to record it? 634 squares (14.1%) *Ranked 56*

A hundred years ago Nuthatches were commonest in the well wooded counties of the southern half of England and in the eastern parts of Wales. They were probably absent from the westernmost parts of Wales and did not breed north of Lancashire and Yorkshire (they had been in Durham 50 years before). A gradual expansion north and west took place over the first 30 years of the 20th century but they then retreated south again. However about 50 years ago the gradual spread north started again and is clearly continuing apace. Between the first and second Breeding Atlases much of lowland Cumbria and Northumbria was colonised and the birds were first proved to breed in Scotland in 1989 and the total has increased to about 20 pairs. There are no records from Ireland. The CBC may have increased three-fold since the mid-1970s and the five-year BBS was up 30%. *All indications are that this attractive species has plenty of suitable woods to colonise in Scotland.*

TREECREEPER (Eurasian Treecreeper)

Certhia familiaris

> *Distribution* Britain 2,120 (-7.3%) Ireland 567 (-20.1%)
> *Numbers breeding*: Britain 200,000 Ireland 45,000
> *European status*: 3,000,000 (8% in Britain and Ireland =5)
> *British population trend*: apparently stable (+13% CBC){-2%}
> *How likely are you to record it?* 558 squares (12.4%) *Ranked* 59 [61]

Treecreepers spread northwards in Scotland during the 19th century and reached North-west Sutherland a few years later. They bred on the Western Isles for a few years in the 1960s and 1970s and again in the 1990s; however they are not known to have bred on Orkney or Shetland. In Ireland they spread in Western Mayo and Connemara fairly recently but there was a considerable loss between the two Breeding Atlases from all over Ireland – there were also losses in Britain. As a very small resident bird they are very vulnerable to cold winter weather. The CBC has fluctuated about a fairly steady mean following the recovery from the 1962/63 winter but, recently, may have increased slightly. *Still vulnerable to winter cold but otherwise set fair.*

SHORT-TOED TREECREEPER

Certhia brachydactyla

This species breeds in the Channel Isles in place of the Treecreeper – which has never been recorded breeding.

PENDULINE TIT (Eurasian Penduline Tit)

Remiz pendulinus

This is a good candidate as a new breeding species with more and more records but the only hint of this happening is of a male nest-building in Kent during April 1990. In Central Europe the birds have spread west from Poland and Slovakia to the far side of the Channel in 50 years. *Coming soon!*

GOLDEN ORIOLE (Eurasian Golden Oriole)

Oriolus oriolus

> *Distribution* Britain 45 (+350%) Ireland 0
> *Numbers breeding*: Britain 24 Ireland 0
> *European status*: 1,400,000 (0% in Britain and Ireland)
> *British population trend:* established but recent decline
> *How likely are you to record it?* 1 squares (0.0%) *Ranked* 190=

The 19th century saw breeding records from many counties – especially Devon and the counties from Lincolnshire to Kent. Many birds were shot as trophies before they could nest. Regular breeding (in Kent) may have ceased by about 1910 but sporadic nesting, mostly in the South-east, continued. However just over 30 years ago the colonisation of poplar plantations in the Fenland basin started. These birds almost certainly came from Holland where the birds prefer these trees for nesting. The population reached 20-35 some 15 years ago and has remained at this level since (maximum 42 in 1990). Despite extensive poplar plantations being felled – the wood was not needed for matches with the introduction of cheap lighters – they spread to breed in smaller areas. The only breeding record in Scotland was during 1974 in Fifeshire but other birds have been recorded, before and since, in summer. *Here to stay and, with the help of global warming, to spread?*
Millwright, R.D.P. 1998 *Bird Study*: **45**, 320-330.

RED-BACKED SHRIKE

Lanius collurio

> *Distribution* Britain 15 (-86.5%) Ireland 0
> *Numbers breeding*: Britain 6 RBBP Ireland 0
> *European status*: 2,900,000 (0% in Britain and Ireland)
> *British population trend:* effectively extinct
> *How likely are you to record it?* 4 squares (0.1%) *Ranked* 167=

The story of the Red-backed Shrike in Britain makes really sad reading. At the end of the 19th century a decline had already been noted with breeding sporadic in England north of a line from the Humber to the Mersey. The bird was still considered common over the rest of England and most of Wales. The retreat in the first 50 years of this century left the birds breeding, less commonly, south of a line from the Wash to South Wales with none left in Cornwall and few in Devon. About 1955 the last one bred in Wales and by 1960 the population was down to about 250. The die was cast and there were about 50 left in the early 1970s (more than half in Suffolk). Sporadic summer records in Scotland, presumably birds from Scandinavia, had been reported but in 1977 three pairs bred and there have been subsequent records there (including proved breeding and other pairs in 1997). Regular breeding in the traditional areas in East Anglia ceased several years ago. This species migrates south-east and we are at the extreme edge of the range. The loss of mixed farms with small fields and large hedges may be a cause of their demise since, where this habitat is still present in Belgium, there is still a good population. However the typical habitat of the last pairs in England was another threatened habitat – lowland heath. The birds were also a target for egg collectors. *Not much chance of a return in the South unless global warming can help.*
UKBAP **Scottish Natural Heritage/English Nature** RSPB.

JAY (Eurasian Jay)

Garrulus glandarius

> *Distribution* Britain 1,713 (-1.8%) Ireland 269 (-32.2%)
> *Numbers breeding*: Britain 160,000 Ireland 10,000
> *European status*: 6,500,000 (3% in Britain and Ireland =10)
> *British population trend*: stable (-1% CBC, -33% farm){-4%}
> *How likely are you to record it?* 1098 squares (24.4%) *Ranked* 44 [71=]

A hundred years ago the Jay was in retreat being persecuted by keepers and for its plumage – both for fly-tying and for the millinery trade. This meant that the breeding birds were confined to the south-east corner of

Ireland and in Scotland had been exterminated from Ayrshire and some other areas in Scotland. Otherwise it bred up to the Highland fringe. The Irish birds, a separate race *hibernicus*, have gradually spread west and north to occupy about two-thirds of the country by the time of the first Breeding Atlas and a bit further by the second. Jays in Ireland are thinly distributed. In Scotland Jays have increased and are expanding north with breeding recently proved in Moray and Nairn and the Great Glen occupied between the two Breeding Atlases. CBC indices have been relatively stable but on farmland there has been a recent decline (also on BBS down 17% in five years). This may be because these are really woodland birds and the intensification of farming has affected them there. *Woodland Jays, at least, seem to be doing well.*

MAGPIE (Black-billed Magpie)

Pica pica

> *Distribution* Britain 1,958 (+1.1%) Ireland 962 (+0.9%)
> *Numbers breeding*: Britain 590,000 Ireland 320,000
> *European status*: 9,000,000 (10% in Britain and Ireland =3)
> *British population trend:* big increase, now stable (+94% CBC){+113%}
> *How likely are you to record it?* 3,079 squares (68.6%) *Ranked* 12 [5=]

Magpies were, and are, just about the most hated of all the crows by gamekeepers and suffered unmercifully at their hands. By the end of the 19th century their populations in much of East Anglia had been severely reduced and the birds were in retreat from northern Scotland. By 1940 it was only common in parts of the Lowlands and its demise elsewhere was

Magpie with Bullfinches.

put down to persecution but this has not been proved. In Ireland the Magpie first arrived about 325 years ago in Wexford and had spread throughout the country – except the far North-west – by 100 years ago. These areas were gradually colonised during the next few decades. In Scotland the birds started to increase about 60 or 70 years ago, possibly through the demise of the keeper and/or because of increased planting of forestry. They have spread up the east coast and now breed in the Black Isle and through the lowland areas of North-east Grampian as well as some extra southern areas. The CBC has shown amazing increases in the 25 years from 1965 with an overall three-fold gain. There is good evidence that this has halted in the last ten years. The great increase coincided with breeding becoming widespread in suburban and urban areas and considerable agitation over nest predation by the Magpies. However detailed research has failed to find any causative link between the demise of many small bird populations and the increase in Magpies. *The outlook seems excellent for a very well adapted species for modern living.*

CHOUGH (Red-billed Chough)

 Pyrrhocorax pyrrhocorax

> *Distribution* Britain 88 (+12.8%) Ireland 168 (+1.8%)
> *Numbers breeding*: Britain 340 Ireland 830
> *European status*: 14,000 (8% in Britain and Ireland = 3=)
> *British population trend*: possibly increasing in some areas
> *How likely are you to record it?* 12 squares (0.3%) *Ranked* 145= [77=]

The Chough is primarily a bird of coastal cliffs and, by the end of the 19th century, had already gone missing from the Lothians, Berwickshire, Northumberland, Kent, Sussex, Hampshire (and Isle of Wight), Somerset, Anglesey, North-west England, the Western Isles and much of the east coast of Ireland. In Ireland this decline carried on for another 25 years but was probably reversed about 1925 and an incomplete 1963 survey found less than 600 pairs, a better survey in 1982 logged up to 685 pairs (only 55 away from the coast). Even better coverage in 1992 showed the major concentrations were in Cork and Kerry (up to 600 pairs) and a possible maximum of over 900 pairs. In Scotland Islay has been the centre of the population for 100 years and three-quarters of about 80 pairs found in 1982 bred there. Since then the population has declined. In Wales the 1992 survey found about 150 pairs – about 30% more than 30 years before – and up to 187 were found in 1996 with increasing numbers inland. The birds are doing well on the Isle of Man – 68 pairs in 1991. Changes to the cliff top ecology diminishing the close cropped sward where the Choughs can feed may quickly cause losses. *These rare birds seem to be relatively safe but there is little sign of lost ground being regained at all quickly.*
Berrow, S.D. *et al.* 1993 *Irish Birds*: 5, 1-10.

JACKDAW (Eurasian Jackdaw)

Corvus monedula

> *Distribution* Britain 2,344 (-3.0%) Ireland 944 (-2.1%)
> *Numbers breeding*: Britain 390,000 + Ireland 210,000
> *European status*: 5,300,000 (11% in Britain and Ireland =2)
> *British population trend*: increasing steadily (+117% CBC){+148%}
> *How likely are you to record it?* 2607 squares (58.0%) *Ranked* 19 [13]

The Jackdaw was a common bird over the whole of Ireland and everywhere in Britain except the North-west of the mainland, the Western Isles (just colonised) and Shetland (no breeding) in 1900. The birds continued to spread and bred on Shetland first in 1943 but it is only sporadic there and there are still very few on Lewis. The numbers have probably increased in Ireland over many years. In North-west Scotland they are still very sparse and there may have been some real losses in this area between the two Breeding Atlases. The CBC index has probably increased three-fold in about 35 years and the five-year (1994-98) BBS index has increased by 13%. Nesting criteria from the NRS are improving too. In areas where pigs are kept out of doors they make a great deal of use of the food fed to the animals. *Outlook very good for a crow that most people seem to like!*

ROOK

Corvus frugilegus

Distribution Britain 2,237 (-0.4%) Ireland 919 (-1.2%)
Numbers breeding: Britain 855,000 Ireland 520,000
European status: 3,500,000 (39% in Britain and Ireland =1)
British population trend: making up losses (+40% survey 1975-96)
How likely are you to record it? 2250 squares (50.1%) *Ranked* 21 [8]

By the end of the 19th century Rooks were almost everywhere that they could find decent trees to nest in. They colonised Orkney in autumn 1846 (bred 1847) and Lewis in October 1893 (breeding from 1895). Shetland was colonised in 1952. The numbers probably increased in Ireland earlier this century. The most dense populations are recorded in the Aberdeen area and through much of Ireland. Some colonies are very large – many thousands of nests – and there is no CBC index. However there was a survey of the UK population in 1975 which indicated a decline possibly connected with pesticide usage and even with Dutch Elm disease. However a repeat survey in 1996 indicated a recovery of 40% and the five-year BBS index (1994-98) increased by 10%. *Good prospects for a species seemingly unaffected by changes in agriculture.*

CROW

Corvus corone

Distribution Britain 2,762 (-0.3%) Ireland 970 (-1.2%)
Numbers breeding: Britain 970,000 Ireland 290,000
European status: 6,200,000 (21% in Britain and Ireland =1)
British population trend: steady increase (+101% CBC){+120%}
How likely are you to record it? 3753 squares (83.6%) *Ranked* 5 [11]

The two races of crow (Carrion *corone* and Hooded *cornix*) and their hybrids bred all over Britain and Ireland from the first records but they were severely persecuted. They were, and are, considered to be very serious predators of game birds wherever they may occur – even on the highest mountain tops where Ptarmigan may be taken. They were also blamed for lamb deaths and so the farmers' guns, traps and poison were also turned against them. A general increase in numbers started with the First World War when so many keepers were re-deployed into the army. This was apparent all over the range but particularly in lowland England and Ireland. In Scotland the hybrid zone between the carrion and hooded forms has moved steadily north and west over the last 70 years. The CBC has increased about four-fold in the last 35 years and the five-year BBS (1994-98) increased significantly by 7%. *Despite legal destruction the Crow is going to be a common bird throughout our area in the future.*
Watson, A. 1996 *Scottish Birds*: **18**, 206-207 (and **19**, 64).

RAVEN (Common Raven)

Corvus   corax

Distribution Britain 1,131 (-9.2%) Ireland 686 (+53.8%)
Numbers breeding: Britain 7,000 + Ireland 3,500
European status: 220,000 (1% in Britain and Ireland =6)
British population trend: declining overall (+37% BBS)
How likely are you to record it? 348 squares (7.7%) *Ranked* 69 [35]

The Raven used to be a bird found throughout Britain but was lost from much of lowland England by the end of the 19th century – for instance breeding in Surrey to about 1850. In Ireland it had been lost from several inland counties but was still found round the coasts and in most hilly areas. As the largest corvid and thought to be the scourge of farm animals and game alike it was high on the list of birds to be destroyed. There was an increase probably starting with the lack of gamekeeping activity during the Second World War in England and rather earlier in Ireland. In Scotland the increase was short lived and a decline set in about 20 years later associated with pesticides, probably sheep dips in carrion, and differences in land use. Losses between the two Breeding Atlases are concentrated in eastern Scotland – all the way from the Borders to Caithness. In Ireland the spread inland continued in many areas and in Wales, where the birds foraged over open sheep grazing, they also increased markedly in numbers. There may be as many as 1,500 pairs breeding there. The five-year BBS index (1994-98) increased by 37%. *Severe losses in Scotland more than counteract increases elsewhere – particularly Wales and Ireland.* Roberts, J.L. & Jones, M.S. 1999 *Welsh Birds*: 2, 121-130.

STARLING (Common Starling)

Sturnus vulgaris

Distribution Britain 2,620 (-3.6%) Ireland 957 (-2.3%)
Numbers breeding: Britain 1,100,000 – Ireland 360,000 +
European status: 40,000,000 (4% in Britain and Ireland = 7=)
British population trend: severe decline (-45% CBC, -77% wood){-58%}
How likely are you to record it? 3,195 squares (71.1%) *Ranked* 10 [15]

This is a species that most people think has been universally common throughout Britain and Ireland since the beginning of recorded time – not so! In Scotland 200 years ago it was only breeding (and there it was common) in the Western Isles, Orkney and Shetland and the very far north-east of Caithness. They increased to cover the whole country by 100 years ago and gradually became more common for the next 50 years. In England and Wales there had been considerable expansion north and west during the 19th century and increases were noted early in the 20th century. These increases were mirrored in Ireland where the far south-east was only fully colonised about 40 years ago. Between the two Breeding Atlases there have been considerable losses from North-west Scotland including some of the Inner Hebrides. These are not easy birds to survey in the CBC but the steady index until about 1980 dipped then and has halved – the five-year BBS index (1994-98) has gone down by 13%. These remarkable losses by such a common species seem to be happening over much of North-west Europe

and are of considerable concern. On intensively farmed land it is thought that the lack of suitable invertebrate food for the chicks is causing poor breeding productivity. *Recent losses are very worrying – is the Starling going to disappear just as quickly as it arrived two centuries ago?*

HOUSE SPARROW

Passer domesticus

Distribution Britain 2,525 (-5.3%) Ireland 901 (-6.8%)
Numbers breeding: Britain 3,600,000 Ireland 1,100,000
European status: 54,000,000 (9% in Britain and Ireland =3)
British population trend: severely declining (-64% CBC){-58%}
How likely are you to record it? 2,701 squares (60.1%) *Ranked* 18 [25]

The House Sparrow was very common wherever people lived by the end of the 19th century, although it had only rather recently reached some parts of the west of Wales and north-west of the Scottish mainland. It was already abundant through the outer islands – although the Outer Hebrides had only been fully colonised about 30 years before. This success was despite the presence of many 'Sparrow Clubs' whose sole object was to rid the parish of the pests! While horses were being used for transport, it must have been really easy living for the birds, feeding from spilt grain, straw bedding, even undigested grain in droppings. They probably declined as thatched roofs were replaced and the human population withdrew from remote areas. This is clearly shown by the losses documented in the second Breeding Atlas – particularly from high areas and from the west coast of Ireland. These are not easy birds to record for the CBC but the 25-year decline was 64% and the BBS five-year index dropped (1994-98) 7%. Many people report that 'their' sparrows have gone and figures provided from Kensington Gardens, from Max Nicholson, are quite amazing and very disquieting:

November 1925	2,603
December 1948	885
November 1966	642
November 1975	544
February 1995	46

It is suspected that the intensification of agriculture is making a difference in the country but, in the towns, it may be the lack of insect food during the vital days when the nestlings need the extra protein. Typically areas occupied by a 'clan' of House Sparrows will lose their birds quite suddenly and this may sometimes be related to the loss of breeding or roosting sites: for

instance, when a building is demolished (or repaired) or an ivy-covered tree cut down or the creeper removed from a wall. *More and more people will be seeing less and less spadgers in the future.*
Summers-Smith, J.D. 1999 *British Wildlife*: 10, 381-386.

TREE SPARROW

 *Passer montanus*

> *Distribution* Britain 1,346 (-19.6%) Ireland 130 (+6.6%)
> *Numbers breeding*: Britain 110,000 Ireland 9,000
> *European status*: 15,000,000 (1% in Britain and Ireland)
> *British population trend:* very severely declining (-87% CBC){-87%}
> *How likely are you to record it?* 277 squares (62.0%) *Ranked* 75 [64=]

Without doubt there have been problems in finding out just what was happening to Tree Sparrow populations a hundred years ago since not everyone reporting on birds realised there were two species of sparrow. However it seems clear that Tree Sparrows were expanding and were found over most of England (not Devon and Cornwall), in Wales (not the West), thinly round coastal Scotland, in the Isle of Man and in Ireland only in Wicklow. The birds continued to increase and spread for about 30 years and several coastal, and a few inland, colonies were established in Ireland. The birds then declined and went missing from many areas. Some 40 years ago they started to increase again and were probably as common as they ever were during the first Breeding Atlas. By the second Atlas the most recent decline had started with the birds lost from many areas from Sussex to South and West Wales and, patchily, from northern England and southern Scotland. In Ireland, there were slight net gains. Tree Sparrows are very fickle and large colonies may be deserted in a very few years. The lack of insects for the nestlings in the modern agricultural landscape is a likely cause. The CBC index dropped by a staggering 87% in the 25 years (1972-96) and mostly in the final 20 years. *This is now a rare bird in many areas and looks set to become even rarer.*
Perry, K.W. & Day, K.R. 1997 *Irish Birds*: 6, 29-34.
UKBAP **MAFF** RSPB.

CHAFFINCH

 Fringilla coelebs

> *Distribution* Britain 2,602 (+0.7%) Ireland 950 (-0.9%)
> *Numbers breeding*: Britain 5,400,000 Ireland 2,100,000
> *European status*: 85,000,000 (9% in Britain and Ireland =3)
> *British population trend:* slow but steady increase (+25% CBC){+30%}
> *How likely are you to record it?* 3,971 squares (88.4%) *Ranked* 2 [5=]

The Chaffinch was a very common and widely distributed species at the end of the 19th century with range and numbers extending and increasing in the far north-west of Scotland and on the Hebrides. Some bred on Orkney but none on Shetland or on Scilly – then not as wooded as now. In upland areas there were increases associated with new plantations and substantial numbers may settle in new areas. Breeding on Shetland is very sporadic. Widespread reports of declines in the late 1950s have been attributed to deaths caused by agro-chemicals but the population seems to have been recovering steadily over the 30 years 1960-90 and the BBS five-year index (1994-98) still shows a significant increase of 4%. The birds are most at home in woodland, but they do not appear to be affected on farmland by the extensive agricultural change that has happened over the last few decades. The birds use beech mast extensively in the winter when it is available and are amongst the most frequent of birds feeding in gardens. *No worries about range or numbers for a familiar species.*

BRAMBLING

Fringilla montifringilla

This finch is a winter visitor to Britain and Ireland and summering records are not uncommon, normally after winters when there have been large numbers. Two summer records have been in Ireland and most of the rest in Scotland and eastern England. Breeding may have taken place a few times before being successful in 1982 (Inverness). Potential breeding records over the 10 years 1988-97 total 24 with a maximum of eight (1995) and one year (1996) with none. *Coming?*

SERIN (European Serin)

Serinus serinus

This tiny finch became more regularly recorded 40 years ago, following expansion in Europe, and first bred in Dorset in 1967 and again in Sussex in 1969. Welcome for a new regular breeding species was premature although up to nine may have bred in any one year – particularly in Devon during the 1980s. Quite a few were in East Anglia and the only reported record in 1996 and 1997 was a male caught in Norfolk on 15th May. Breeding started on Jersey in 1972 and there may be 20 or more pairs now. *May yet become established on the mainland.*

GREENFINCH (European Greenfinch)

Carduelis chloris

> *Distribution* Britain 2,323 (-2.7%) Ireland 813 (-11.7%)
> *Numbers breeding*: Britain 530,000 Ireland 160,000
> *European status*: 13,000,000 (5% in Britain and Ireland = 5=)
> *British population trend*: stable (+3% CBC, +19% farm){-2%}
> *How likely are you to record it?* 2,828 squares (63.0%) *Ranked* 15 [19]

By the end of the 19th century this was a common and widespread species throughout Britain save for the very North-west and the Hebrides. Orkney had recently been colonised but they were not on Shetland. In Ireland they were only missing from the bleak and treeless areas of the West coast. The birds continued to spread in Scotland and Western Ireland helped by new plantations and soon reached Lewis and many other Hebridean islands. Breeding is still not recorded on Shetland and the outer islands of the Orkneys – nor in parts of the North-west Mainland of Scotland. The Greenfinch has become a very widespread and common feeder in gardens and this has probably insulated it against potential losses due to agricultural intensification. There were some losses between the two Breeding Atlases particularly in the northern half of Ireland and in upland areas in Scotland and Wales. The CBC index has stood up well (overall 3% increase 1972-96) and recent BBS indices over the five years 1994-98 show a healthy and significant 13% increase. However the NRS shows a worrying recent (last 15 years) increase in nest failures. *This is one finch species that looks as if it might be coping rather well with the modern world.*

GOLDFINCH (European Goldfinch)

Carduelis carduelis

> *Distribution* Britain 2,209 (+5.4%) Ireland 749 (-17.9%)
> *Numbers breeding*: Britain 220,000 Ireland 55,000
> *European status*: 8,000,000 (3% in Britain and Ireland = 9=)
> *British population trend*: stable (+2% CBC, +18% farm){+10%}
> *How likely are you to record it?* 2,175 squares (48.4%) *Ranked* 25 [29]

Towards the end of the 19th century the Goldfinch was in retreat and severely depleted in many areas through the catching of large numbers of birds for caging and, possibly, through agricultural change. The birds were no longer breeding in Northern Scotland, were rare in the South and through much of Northern England and near large towns in Ireland – Wales seemed largely unaffected. Protection laws began to have their effect and the agricultural depression increased the weedy areas where they feed and so there was a reasonably steady increase until the early 1960s. Then the severe winters caused a sharp drop followed by a rather slow recovery. In about 1975 numbers started to drop (CBC) by over 40% during the next ten years, possibly due to the use of more efficient herbicides, but the birds have largely recovered since. Between the two Breeding Atlases there were significant gains from Aberdeen to the North coast of mainland Scotland although they still do not breed on Orkney, Shetland and the Outer Hebrides. In Ireland the was a net loss of over 150 10-km squares, but, in some areas, they now breed in mature conifer plantations. *Set fair for one of our most brightly coloured birds.*

SISKIN (Eurasian Siskin)

Carduelis spinus

> *Distribution* Britain 1,158 (+85.3%) Ireland 284 (+19.8%)
> *Numbers breeding*: Britain 300,000 Ireland 60,000
> *European status*: 2,750,000 (13% in Britain and Ireland =3)
> *British population trend*: long-term increases (+12% BBS)
> *How likely are you to record it?* 272 squares (6.1%) *Ranked* 76 [67]

A mere two hundred years ago the Siskin was a bird only found in the Caledonian pine forest since it relies on mature cones for food. By the end of the 19th century it had spread a little to the maturing conifer plantations of other areas – North Wales, southern Scotland and possibly Cumbria. In Ireland it was apparently well distributed following an invasion in the 1850s. Breeding records in Surrey are thought to have been from escaped cage birds. Many conifers were planted in the early part of the 20th century and as they matured, in the 1930s, breeding birds spread to new areas and, by the 1950s, they were breeding as far south and east as Devon, Hampshire, Kent and Suffolk. They had consolidated their breeding numbers and expanded their range in Ireland too. However even by the first Breeding Atlas nesting can only be described as sparse away from parts of Scotland and favoured areas in Ireland. The second Breeding Atlas showed tremendous expansion and real filling in of the range with an expansion of two thirds, overall, in the number of occupied 10-km squares. This bird is very fickle over its breeding areas and will leave places where it generally breeds if the cone crop fails – but pile in if good. *Bound to fluctuate but all in all on the up and up!*

LINNET (Common Linnet)

Carduelis cannabina

> *Distribution* Britain 2,268 (-4.6%) Ireland 783 (-17.9%)
> *Numbers breeding*: Britain 520,000 – Ireland 130,000
> *European status*: 8,000,000 (8% in Britain and Ireland =4)
> *British population trend*: seriously declining (-41% CBC){-38%}
> *How likely are you to record it?* 2,218 squares (49.4%) *Ranked* 23 [27]

The Linnet may have lost out to the growth of agriculture as scrubby areas with gorse bushes were cultivated during the 19th century, and to the depredations of bird catchers near large population centres. By the end of the century they were present over most of Britain and Ireland except the Outer Hebrides and Shetland with sparser populations in North-west Scotland and the Inner Hebrides. The birds may have expanded a little during the first half of the 20th century – breeding may have taken place even on the Western Isles and Shetland for a time. By the time of the first Breeding Atlas there may have been losses from North-west Scotland and there were many losses from Scotland and Ireland by the time of the second Atlas. This is a species that requires weed seeds and has suffered as the technology of herbicides has been developed. The change on the farmland CBC, 1972-96, has been a decline of 41% and the only grain of comfort is that it has been relatively stable for more than ten years, possibly because they like oilseed rape. *At the moment no sign of a return to earlier (proper?) levels.* UKBAP **MAFF** RSPB.

TWITE

Carduelis flavirostris

> *Distribution* Britain 651 (-1.1%) Ireland 60 (-52.8%)
> *Numbers breeding*: Britain 65,000 Ireland 3,500
> *European status*: 300,000 (24% in Britain and Ireland =2)
> *British population trend*: probably declining
> *How likely are you to record it?* 47 squares (1.0%) *Ranked* 115 [88=]

The mountain version of the Linnet, this species was (and probably is) much confused with it. However at the end of the 19th century the bird was widely distributed through all Scotland, through northern England and south to Derbyshire, the Isle of Man and much of Ireland. It was absent from most lowland farming areas preferring rocky and upland sites and was particularly common around the coasts of Northern Scotland

and on the Outer Isles. In Ireland there seems to have been a general decline and contraction in range over the last 80 years with 50% reduction in 10-km squares occupied between the two Breeding Atlases. It is now almost wholly confined to the West and North coasts. In Britain the birds retreated northwards and gradually became restricted, when at all numerous, to north and west of the Great Glen. For a time none bred in the southern uplands of Scotland although there were some left on the Solway shore. In England they were in Devon for a short while but were lost from the Isle of Man and became restricted to the Southern Pennines and the peak District. However the second Breeding Atlas showed some improvement but breeding densities in the core area have decreased and nest failures (NRS) have increased. Detailed studies show that they lose out when heather moorland becomes grass dominated. *The losses in Ireland are particularly worrying but they may be just about holding their own elsewhere.*
Brown, A.F., Crick, H.Q.P. & Stillman, R.A. 1995 *Bird Study* **42**, 107-121.

REDPOLL (Common Redpoll)

Carduelis flammea

Distribution Britain 1,754 (-11.3%) Ireland 538 (-35.9%)
Numbers breeding: Britain 160,000 – Ireland 70,000
European status: 1,500,000 (15% in Britain and Ireland =4)
British population trend: widespread serious declines (-89% CBC){-92%}
How likely are you to record it? 286 squares (6.4%) Ranked 74 [50=]

A hundred years ago the characteristic calls of the Redpoll were heard in summer in most areas of Britain and Ireland. The race breeding with us is *cabaret* but a very few Continental *flammea* have been recorded breeding – mostly in northern Scotland. In Ireland it was still spreading west but seems to have been fairly common. In England the birds were rare south of a line from the Wash to the Severn. They were more common further north and in the upland areas of Wales. They were absent from the Outer, and many of the Inner, Hebrides, Shetland and probably Orkney. The expansion continued for a time but the species started to decline from 1920 for about 30 years when a new expansion started. By the time of the first Breeding Atlas a few bred in Cornwall, Orkney and the Outer Hebrides but the was a very obvious gap, about 100 kilometres square, from the New Forest northwards. Huge losses were recorded in the second Breeding Atlas particularly in south and East Ireland, North-east Scotland and the midlands of England. The CBC index, having increased five-fold in the late 1960s and early 1970s, plummeted by 89% (1972-96)! *Redpolls are still around and have increased rapidly in the past – let's hope they do it again.*

CROSSBILL (Common Crossbill)

 Loxia curvirostra

> *Distribution* Britain 763 (=153%) Ireland 156 (=5100%)
> *Numbers breeding*: Britain 10,000 ~ Ireland 1,000 ~
> *European status*: 1,200,000 (0% in Britain and Ireland)
> *British population trend*: wildly fluctuating but prospects good
> *How likely are you to record it?* 110 squares (2.4%) *Ranked* 101 [101=]

The crossbills are a group of species whose fortunes follow the cropping of their food trees: the conifers. Their numbers and breeding areas change grossly from year to year and the three species we have in Britain and Ireland are difficult to tell apart. This one is the commonest and may breed almost anywhere after an irruption, when flocks reach us from the Continent, but they do need mature conifers! This means that there were far fewer breeding, even after irruptions, 100 years ago. Then Caledonian forest and some early conifer plantations in Scotland and the first Breckland plantations may have had regular populations. It seems likely that several areas in Ireland had breeding birds for many years following irruptions in the 1860s. Irish records diminished, even following irruptions, and there were no breeding records in the first Breeding Atlas (present in three 10-km squares) – but there were records from 156 squares in the second following big irruptions. Now there are many more mature conifers and Britain had records from many areas for the first Breeding Atlas and far more for the second. *Britain and Ireland have many more mature conifers now than at any other time and are bound to be a good reception area for wanderers from the Continent.*

SCOTTISH CROSSBILL

 Loxia   scotica

Distribution Britain 59 (??%) Ireland 0
Numbers breeding: Britain 1000 ? Ireland 0
European status: 600 (100% in Britain and Ireland =1)
British population trend: steady, we hope!
How likely are you to record it? 7 squares (0.2%) *Ranked* 159

These birds are not at all easy to identify but they are thought to be long-term residents of the Caledonian pine forests and the only species endemic to Britain. They were identified from 59 10-km squares in the second Breeding Atlas (not recognised as separate species at the time of the first). Recently it has been suggested that some of the bigger billed birds breeding in these areas may be irrupting Parrot Crossbills. *The plot thickens but, if they exist, the Scottish Crossbill seems to be secure.*
UKBAP **Scottish Natural Heritage** RSPB

PARROT CROSSBILL

 Loxia   pytyopsittacus

These bigger billed birds sometimes irrupt to Britain and were first recorded breeding in Norfolk in 1984 and 1985 and then in Scotland and Breckland in 1991. Display has now been seen elsewhere and pockets of breeding birds of this species may already be established.

COMMON ROSEFINCH (Scarlet Rosefinch)

Carpodacus   erythrinus

This is a species that has galloped across Europe colonising new countries over the last century. First breeding (Scotland) was in 1982 and again in 1990. There were then up to eight pairs breeding in Yorkshire and Suffolk during 1992 but this did not indicate they had *really* arrived – one more pair in 1997 is all we have had proved so far. There was a singing male in Co. Mayo in 1998. *Ten or more singing males in a year regularly – when the females come they will become properly established.*

BULLFINCH (Common Bullfinch)

Pyrrhula pyrrhula

> *Distribution* Britain 2,173 (-6.5%) Ireland 825 (-5.1%)
> *Numbers breeding*: Britain 190,000 — Ireland 100,000
> *European status*: 3,000,000 (9% in Britain and Ireland =4)
> *British population trend*: seriously declining (-62% CBC){-40%}
> *How likely are you to record it?* 988 squares (22.0%) *Ranked* 49 [28]

The Bullfinch was much prized as a cage bird in the 19th century but protection laws may have helped it to increase by 1900. Then it was found all over Britain except the far north of Scotland and the outer Isles and Isle of Man. In Ireland it was increasing from the East and was just about breeding throughout the country. There were local increases, especially where there were new plantations, until 1950 or so and it then started to become apparent that the birds were also colonising more open habitats possibly as Sparrowhawk predation declined. The Bullfinch is still absent from the same areas and suffered net losses (6%) between the two Breeding Atlases. However the worst news was the 62% decline in the 25-year (1972-96) CBC and even 27% in the five year (1994-98) BBS index. Loss of the straggling hedges they prefer and the general intensification of agriculture could be the cause and, if lack of it is implicated in the earlier increase, the renewed predation by Sparrowhawks might affect the index values. *No sign, yet, of an end to the downward slide – will it ever end!*

HAWFINCH

Coccothraustes coccothraustes

> *Distribution* Britain 315 (-31.4%) Ireland 0
> *Numbers breeding*: Britain 4,750 Ireland 0
> *European status*: 1,300,000 (0% in Britain and Ireland)
> *British population trend*: apparently declining badly
> *How likely are you to record it?* 3 squares (0.1%) *Ranked* 171=

This spectacular large and big-billed finch is not easy to find and record. It may have bred very sparingly 200 years ago but, by the end of the 19th century, it bred through most of England (except the south-west peninsula), Wales (except for the West) and just possible in Midlothian. By 1920 it was breeding in Dumfries and more widely in the eastern lowlands. Later some were found breeding near Aberdeen and the birds gradually completed the colonisation of England and Wales but did not reach Ireland to breed. Possibly they reached a peak of numbers and range at the time of the first Breeding Atlas for they were recorded from 31% less 10-km squares in the second. Several detailed series of records show that the birds have really declined and that this is unlikely to be a result of the different recording methods. A pair may have bred in Co. Clare in 1991. The birds are not yet regular garden bird feeders here but are in many areas on the Continent (e.g. Vienna). *If they learn to do this they may regain their former numbers and distribution.*

LAPLAND BUNTING (Lapland Longspur)

Calcarius lapponicus

Following a late, cold, spring in 1977, when there were a lot of spring reports, breeding probably took place at five sites (up to 16 pairs) and there were up to six in 1978 and in 1979 up to 14. They then fizzled out with one pair in 1980 and only one bird ion 1981 – just like 1968 and 1974. Since then single birds have been recorded in 1989, 1995 and 1997 (twice). *Was the burst of breeding just a tease?*

SNOW BUNTING

Plectrophenax nivalis

> *Distribution* Britain 42 (200%) Ireland 0
> *Numbers breeding*: Britain 85 Ireland 0
> *European status*: 350,000 (0% in Britain and Ireland)
> *British population trend*: may be stable

Snow Buntings were first proved to breed at the end of the 19th century and fluctuate in numbers from year to year. A few singing birds (and even breeding records) come from Orkney and Shetland (and even St Kilda) but the real centre of the population is the tops of the mountains – and particularly Cairngorm. An excess of males is often reported and they are of both the *nivalis* and *insulae* races, but more of the latter. Recently it has been realised that there may be as many as 70 to 100 pairs in a good year rather than the handful of pairs previously thought to be the norm. *Global warming may banish the bird's habitat from Scotland but, otherwise, they seem secure.*

Smith, R.D. 1996 *Ringing & Migration*: **17**, 123-136.

YELLOWHAMMER

Emberiza citrinella

> *Distribution* Britain 2,224 (-8.6%) Ireland 587 (-37.4%)
> *Numbers breeding*: Britain 1,200,000 – Ireland 200,000
> *European status*: 19,000,000 (7% in Britain and Ireland =5)
> *British population trend*: recent severe declines (-60% CBC){-43%}
> *How likely are you to record it?* 2,225 squares (49.5%) *Ranked* 22 [31]

At the end of the 19th century Yellowhammers bred commonly throughout mainland Britain and Ireland, in some numbers on Orkney and sporadically on the Outer Hebrides but not on Shetland. They have always been rather uncommon in areas devoid of cover. In Ireland they had begun to withdraw from the west and north by about 1940 but there were few gaps in the first Breeding Atlas. Just 20 years later, with the second Atlas, there were huge losses (over 37% of 10-km squares lost) with great swathes of the country missing out on these once common birds. In Scotland breeding ceased on Orkney and in the Western Isles 30 years ago and the birds have been lost from many west coast and upland 10-km squares, as well as upland areas in North-west England and in Wales. Until the late 1980s the CBC index for Yellowhammer seemed very stable but, in the last ten years, it has dropped like a stone – by about 60% overall. The five-year BBS (1994-98) also went down by 16%. *Another species in dire straits and still falling: out of control?*

CIRL BUNTING

Emberiza cirlus

> *Distribution* Britain 29 (-83.2%) Ireland 0
> *Numbers breeding*: Britain 380 RBBP Ireland 0
> *European status*: 1,600,000 (0% in Britain and Ireland)
> *British population trend*: increasing after severe decline
> *How likely are you to record it?* 6 squares (0.1%) *Ranked* 160=

To most bird watchers the Cirl Bunting is a rare bird restricted to part of South Devon but this was not the case a hundred years ago when it was common throughout the counties south of the Thames, local in the Midlands and breeding on parts of some counties north to North Wales and even in Yorkshire. There may have been as many as 10,000 breeding pairs. The collapse southwards began in about 1930 and by the first Breeding Atlas they were very patchy and recorded from 173 10-km

squares and, by the second, 29! There were probably 200 pairs, or fewer, in 1982 and less than 150 in 1989. Following RSPB research, special measures were taken with farmers encouraged, through stewardship schemes, to provide weed rich stubbles in the core area of South Devon. This has small fields, mixed farming and very good hedges. The population has responded immediately and was estimated at 380 pairs in 1997. *The species now seems safe in its refuge area but will it be able to expand? Possibly.*
Evans, A. 1997 *British Birds*: 90, 267-282.
UKBAP MAFF RSPB & English Nature.

REED BUNTING

Emberiza schoeniclus

> *Distribution* Britain 2,188 (-11.7%) Ireland 831 (-12.3%)
> *Numbers breeding*: Britain 220,000 — Ireland 130,000
> *European status*: 3,600,000 (10% in Britain and Ireland =5)
> *British population trend*: severe declines (-64% CBC){-52%}
> *How likely are you to record it?* 682 squares (15.2%) *Ranked 55 [32]*

At the end of the 19th century this was a very widespread and common bird of damp and wet habitats throughout Britain and Ireland but absent from Shetland, very patchy on the Western Isles but had colonised Orkney about 50 years earlier. Drainage of marshes and other wet areas will have restricted its breeding space but this may well have been balanced by the provision of new reservoirs, gravel pits, etc. About 70 years ago it was realised that the birds were starting to colonise dryer habitats including grassland, scrub and plantations as well as arable areas. The birds seem to have spread in Scotland to reach virtually all areas – first breeding on Shetland in 1948. The first Breeding Atlas might record their maximum spread as, by the second, they were lost from almost 12% of the 10-km squares they had occupied. The southern third of Ireland, the upland areas of Scotland and Wales and the

South-west of England lost a lot of birds. CBC losses were particularly noticeable around 1980 and recently – 64% down over the period between 1972 and 1996. It seems that declining survival rates of the fledged birds may be responsible. *No sign of any recovery but still present in most wetland sites.* Peach, W.J., Siriwardena, G.M. & Gregory, R.D. 1999 *J. Appl. Ecol.*: **36**, 798-811. UKBAP **English Nature** RSPB

CORN BUNTING

 *Milaria calandra*

Distribution Britain 921 (-32.1%) Ireland 11 (-83.8%) *Numbers breeding*: Britain 19,800 Ireland <30 *European status*: 4,600,000 (0% in Britain and Ireland) *British population trend*: severe decline (-74% CBC){-85%} *How likely are you to record it?* 308 squares (6.9%) *Ranked* 71

The Corn Bunting was very widely distributed at the end of the 19th century but so dependant on arable farming that it did not penetrate far, if at all, into upland areas. It does not depend on trees or bushes and, presumably because of the lack of competition, was particularly numerous in the outlying farmland of, for instance, the Outer Hebrides. The decline in cereal growing from 1870 to 1930 was probably responsible for the losses recorded in many areas. There was a recovery for 20 or 30 years but then a further, steeper, decline set in. In Ireland the population had always been most plentiful on the coast but by 1950 they were confined to 'cultivated headlands and coasts'. Huge gaps had appeared in the distribution by the first Breeding Atlas and by the second the occupied 10-km squares had been reduced by 32% in Britain and in Ireland by almost 84%! They may now be extinct in Ireland. There may be a few pairs left in North-east Wales. In Scotland the Uists and Tiree, the Solway and the East coast are almost the only areas still occupied. In England the losses are concentrated in the North and South. The 25-year CBC decline (1972-96) was 74% and the five-year BBS (1994-98) was 42%! The lack of seed-rich winter stubbles and spring-sown barley have been implicated along with the use of pesticides and reduction of mixed farming. *One has to seriously wonder whether this inoffensive and unspectacular bird will survive in Britain.*
Donald, P.F., Wilson, J.D. & Shepherd, M. 1994 *British Birds*: **87**, 106-131.
Donald, P.F. *et al.* 1996 *Scottish Birds*: **18**, 170-181.
UKBAP **MAFF** RSPB & English Nature.

PERSONAL POSTSCRIPT

It is impossible to be immersed in the fortunes of British and Irish breeding birds for a year without wondering what species are doing really well and what really badly. So I will end up with my personal listings of the winners and losers over the 20th century – separating the new colonists and birds which have recovered, in a big way, during this period. These are rather biased towards the birds that are doing well and there are clearly other species equally deserving of listing to those chosen. I have tried to choose one or two species from several categories rather than five or six from a single group. For instance half a dozen could have appeared as successful from the seabirds or raptors or all those failing could have been farmland birds.

It is not fair to use my 'smiley faces' assessments to gauge how well our birds are doing overall as the species that seem to be 'set fair', but not increasing, will be assessed as a single smile and only the few – where there are both optimistic and pessimistic prognoses – are given a 'flat face'. In summary, I was rather amazed that there were 32 more species with good prospects rather than bad! The four categories are summarised as:

Double smile	39	Double scowl	30
Single smile	79	Single scowl	56

I blame the RSPB and English Nature, together with the Wildlife Trusts and other organisations, for doing so very well. Basically over the last forty years rare birds and their special habitats have been very well protected. These rare sites have been the prime target of organisations devoted to running nature reserves. The reserves may be designated for the birds themselves or for plants, landscape value – even a rare spider – but the effect is to help the birds that need the habitat. As some habitats, such as flower-rich meadows or lowland heaths, have been destroyed, the fragmented remains become much more important – and valued. It is the very widespread habitats (and species) that are suffering the worst. Who would have thought, a few decades ago, that spring sown cereals and winter stubbles would have been elbowed out by autumn sowing any more than that hay meadows, where Corncrakes were widespread a hundred years ago, would be replaced by silage cut much earlier? Incidentally the Corncrake is a species that could have been in the losers list or in the 'turned round' species as the numbers are now creeping up due to good research and a partnership with the crofters and other farmers.

The take home message from this book is that several groups of birds are generally doing well – like the birds of prey and seabirds – and many other species that can be looked after in special reserves are also in good shape. However we still have grave problems with many of the once familiar birds of our farmland – species like Corn Bunting, Turtle Dove, Grey Partridge, Spotted Flycatcher and especially the Lapwing. All these will be helped by the huge increase in CAP (Common Agricultural Policy) funding for agri-environment schemes announced in December 1999. The sum involved (£1.6 billion) dwarfs the current spending on nature conservation and will, if used properly, make a tremendous difference both to the environment and the hard-pressed smaller farmer. However I have to record the verdict covering the whole of the 20th century as *could have done better* – and let's hope that we will do so in the 21st.

THE CENTURY'S TOP TEN WINNERS

These species have done really well in the last 100 years and seem set to continue to prosper:

1 RED KITE
Down to one breeding female in mid-1930s but gradually built up in Central Wales and, more recently in England and Scotland through re-introductions. Now 300 pairs and 2,500 by 2010?

2 FULMAR
A hundred years ago this was a rather rare seabird breeding only on remote St Kilda. After taking to feeding on fish discards at trawlers it spread to breed around all our coasts – now 570,000 pairs.

3 BLACKCAP
This sweet-singing wood and scrub warbler expanding in range and numbers, particularly in last 50 years and in northern areas and Ireland. Population now 620,000 pairs in Britain and Ireland and still increasing.

4 GREAT SKUA
Rare large gull-like piratical seabird that only had a small foothold in Scotland in 1900. Now about 8,500 breed there (mostly Shetland) and this is more than half the World population!

5 HOBBY

This elegant little falcon is our only widespread migrant bird of prey. A hundred years ago a few bred in Southern England. Over the last 30 or 40 years they have spread north and increased and, possibly due to climate change and/or reduced persecution, they have expanded to over 600, mostly on farmland!

6 MAGPIE

The familiar black-and-white crow was much diminished at the end of the 19th century but they have increased hugely over the last 40 years with about 910,000 pairs now breeding – possibly up three-fold.

7 SISKIN

This is the tiny greenish finch that breeds in pines and was only really found in the ancient Caledonian pine forests of Scotland until conifers started to be planted for forestry. Now common as a breeding bird in many areas – 360,000 pairs.

8 GANNET

Our biggest seabird – white with black wing tips – breeding colonially on remote headlands and distant islands. About 40,000 or 50,000 pairs bred 100 years ago but there are now five times that number with many new colonies.

9 NUTHATCH

Charming dagger-billed woodland bird with blue back and head. It used to be common only in Southern England but has expanded to reach Scotland about ten years ago and colonised all of Wales – probably helped by global warming – 130,000 pairs.

10 GUILLEMOT

Very common colonial seabird doing very well because over-fishing for large fish has resulted in the seas having large numbers of the small fish it needs. Numbers increasing rapidly and may be nearer to 2,000,000 than the latest estimate of 1,200,000 individuals counted about fifteen years ago. Even the tens of thousands of birds, from our breeding colonies, killed by the *Erika* oil will make little difference to the main populations although it may wipe out the small local colonies.

Other strong candidates included Tufted Duck, Goosander, Honey BuzzardBlack-headed Gull, Reed Warbler and Crossbill – look at their species entries. Remember that species that have newly invaded have their own category.

BOTTOM TEN LOSERS IN BRITAIN AND IRELAND

These birds have dwindled, or crashed, in numbers over the last 100 years and their prospects are poor.

1 RED-BACKED SHRIKE
Common a hundred years ago over much of England and into Wales, this strikingly beautiful small migrant was down to 250 pairs by 1960. It was virtually extinct by 1990, with the last birds breeding on heathland in East Anglia.

2 LAPWING
At the beginning of the century Lapwings (or Peewits) were so common that their eggs were collected for the table. Now they seem to be in free fall with a decline of 49% in 11 years in England and Wales with current estimates as low as 126,300 pairs (certainly no more than 140,000) for the whole of Britain.

3 BLACK GROUSE
This spectacular woodland grouse used to breed in the Home Counties and has been lost from huge tracts of Britain over the last 100 years – its bigger cousin the Capercaillie is doing even worse. Habitat destruction seems to have been the main cause. Population likely to be 15,000 with only 6,500 males lekking. Caper may be now be less than 1,000 individuals – halved in five years! If you wish to watch them, not at an organised reserve, get a leaflet from the RSPB or look on their website (The Lodge, Sandy, Beds SG19 2DL or www.rspb.org.uk).

4 CORN BUNTING
This is perceived as a boring little brown bird and has been forced out by the intensification of agriculture – possibly extinct in Ireland and in awful trouble in Britain with total population estimated at less than 20,000 pairs. Other finches and buntings have suffered too – like Linnet, Redpoll, Yellowhammer and Reed Bunting.

5 TURTLE DOVE
This lovely little migrant dove has retreated from the North and West of its breeding range and seems now to be in free fall – estimate of 75,000 pairs may now be much too high. Recent research suggests that the birds cannot breed for nearly as long as they used to – because of modern intensive farming.

6 SNIPE
This long billed marshland bird used to breed all over the country, but

not now! Drainage of small mashes and wetlands excludes it from breeding over vast tracts of lowland Britain and numbers are also going down in moorland areas. About 65,000 pairs left.

7 GREY PARTRIDGE

This once common native gamebird is almost extinct in Ireland and completely lost from many areas in Britain through agricultural intensification. These are totally unnecessary losses since where they are important for shooting, and the habitat is looked after, they are still doing well! Population less than 150,000 pairs.

8 SKYLARK

Millions less now breed in Britain and Ireland as their agricultural habitat becomes less and less hospitable both in summer and winter. The change from spring to autumn sowing has been disastrous – less breeding sites and less winter food. Nonetheless they are still common – probably just over 1,000,000 pairs in Britain. But the loss of this common bird is mirrored by losses to the House and Tree Sparrows.

9 WRYNECK

This brown migrant woodpecker used to breed over much of Britain but started to decline at the beginning of the century and retreated to Kent and Suffolk. It is now effectively extinct – although a pair or two sometimes breed in Scotland as overspill from Norway.

10 SPOTTED FLYCATCHER

This once familiar garden and woodland bird flies to the South of Africa for the winter. Over the last thirty years some 80% of the breeding population has been lost and the declines seem set to continue. Possibly less than 100,000 pairs in Britain and Ireland now.

Many other candidates included the other species mentioned above and Leach's Petrel, Swift, Lesser Spotted Woodpecker, Yellow Wagtail, Nightingale, Song Thrush, Willow Tit and Bullfinch.

THE FIVE MOST SUCCESSFUL NEW COLONISTS

1 COLLARED DOVE

An urban version of the Turtle Dove – but a resident – these birds first bred in Norfolk in 1955 and there are now almost 250,000 pairs breeding throughout Britain and Ireland: an unbeatable record!

2 GOLDENEYE

This hole nesting diving duck started to breed in nest boxes in Scotland regularly 30 years ago. It has spread out from the initial sites and there are now well over 200 pairs.

3 AVOCET
The black and white wader, the RSPB logo, started to breed on flooded land during the Second World War, where sea defences were breached to defend against invasion. Now nesting in dozens of colonies and with a total of over 600 pairs.

4 CETTI'S WARBLER
Noisy resident warbler of wet scrubland that colonised Britain in 1961. It has now bred in more than twenty counties in England and Wales and there are over 500 pairs.

5 LITTLE EGRET
Elegant small white heron that has colonised Southern England and Southern Ireland within the last four years. Seems set to take off unless we have very cold winters. There are now several dozen pairs in a handful of colonies where they breed alongside Grey Herons.

Runners up in this category include Little Ringed Plover, Black-tailed Godwit, Mediterranean Gull, Black Redstart, the tiny Firecrest and Golden Oriole.

FIVE SPECIES THAT CAME BACK FROM THE BRINK

1 PEREGRINE
With the introduction of organo-chlorine chemicals this big falcon was severely affected in the late 1950s – possibly on the way to extinction. Strict bans on the chemicals have led to an excellent recovery and there are now more than ever breeding – over 1,550 pairs across Britain and Ireland.

2 CIRL BUNTING
Very similar to and a close relative of the Yellowhammer this was a very common bird in Southern England and occurred in the Midlands and Wales. It retreated rapidly to a small enclave in South Devon where careful habitat restoration has more than trebled the population to 380 pairs over the last ten years.

3 SPARROWHAWK
Just about the commonest bird of prey, agricultural chemicals reduced the population in many areas – wiping them out in some. Bans on the compounds have led to a remarkable recovery and there may now be a healthy population of 43,000 or so pairs in Britain and Ireland.

4 WOODLARK
Reduced from breeding through most of England and Wales to about 100 pairs 40 years ago, this species has recovered and expanded. Much

of this has been due to research leading to good habitat management in forestry areas. Population now well over 1,500 pairs.

5 WHIMBREL

A hundred years ago this smaller version of the familiar Curlew was about to be lost from its remote northern breeding grounds in Shetland where there were only about 30 pairs. There has been a strong recovery and gradual range extension with well over 500 pairs now breeding.

Other candidates included Dartford Warbler, Great Crested Grebe, Osprey, Nightjar and Marsh Harrier. Red Kite would have headed this category if it were not already chosen as outright top of the winners.

Finally there are two contentious issues – one a series of potential disasters and the other a delicious series of predictions. The first concerns the possible new introductions intentionally or by mistake by man and the future of the species already introduced. I hope that the attempts to eradicate Ruddy Ducks succeed but fear they will not, and I see the further spread of Egyptian and Canada Geese as very likely. These are three species of wildfowl which have not co-evolved with our avifauna and cause real problems. The impact of species such as Red-legged Partridge and Little Owl are no longer of concern and they, in any case, are from close to us. Of other species the parakeets have yet to exhibit any problems – but they may in the future – and Mandarins would seem to be an ornament of our woodlands. There are, however, real problems lurking just across the North Sea where the House Crow *Corvus superbus* is breeding in the Netherlands. These birds came as ship assisted colonists and they could be a very serious problem, as they have proved in many ports in Asia and Africa. One ship assisted bird, a curiosity: two ship assisted birds, potential disaster. They are very clever nest predators and very adept at living commensally with man. We certainly do not want them to settle and breed or the biodiversity of our native birds could suffer severely.

Natural colonists are to be welcomed. There is one that may be here already – Swinhoe's Petrel – but the others are not yet fully established. They include, in systematic order, several species that appear in the list: Spoonbill, Black Tern, Hoopoe, Bluethroat, Great Reed Warbler, Icterine Warbler, Penduline Tit and Common Rosefinch. But there are others never suspected yet of breeding and even not recorded in the wild. The huge Black Woodpecker *Dryocopus martius* is a strong candidate and the tiny Fan-tailed Warbler (or Zitting Cisticola) *Cisticola juncidus* could take advantage of global warming. Coming across into Northern Europe is the Greenish Warbler *Phylloscopus trochiloides* and a close relative that could come up through France is the Western Bonelli's Warbler *Phylloscopus bonelli*. Perhaps the longest shot of all, never accepted in the wild in Western Europe is the Long-tailed Rosefinch *Uragus sibiricus* currently on the fringe of the very South-east corner of the Western Palearctic and apparently rapidly colonising new areas every year.

The 21st century will undoubtedly be interesting for our birds. There will be more species as the taxonomy modernises and the DNA scientists find out just how far apart, from the evolutionary point of view, are some populations which we now consider to be the same. For instance Carrion and Hooded Crow are finally going to be awarded specific status in the near future – and there are several hundred thousand of each breeding in Britain and Ireland. However the real struggle for conservation will be to maximise the biodiversity we can retain in the ordinary countryside. As Ron Murton said about thirty years ago 'Conservation in a few years time will not be about Ospreys and other rare birds in reserves but about whether someone walking their dog in Croydon can see and hear a Song Thrush.' In a hundred years' time I am sure that this is how our efforts over the next century will be judged.

Lesser Whitethroat.

BIBLIOGRAPHY

The books, journals and papers listed here are used, where appropriate, throughout the book. Most are described in more detail than would normally be the case but several are so important that readers of this book may want to read or purchase them themselves. Items used for particular chapters are listed at the end of the chapters, and papers and reports used for particular species entries – over and above the ones listed here – appear at the end of each species entry. No references are made to county bird books or reports, or to publications substantially pre-dating the Second Breeding Bird Atlas. No attempt has been made to plumb the murky waters of the flurry of taxonomic changes about to be launched on the birding world. The bold italic names at the beginning of each entry is how the item is generally referred to in the text.

The Second Breeding Bird Atlas. Gibbons, D.W., Reid, J.R. & Chapman. R.A. 1993 **The New Atlas of Breeding Birds in Britain and Ireland: 1988-1991.** *T & AD Poyser.* The vast amount of field work put in by members of the BTO, SOC and IWC provided the raw material for this fascinating book full of information and maps – essential!

The First Breeding Bird Atlas. Sharrock, J.T.R. 1976 **The Atlas of Breeding Birds in Britain and Ireland.** *T & AD Poyser.* Superseded by the second Atlas the first one provided the bench-mark detailed maps against which all subsequent ones will be assessed.

The EBCC Bird Atlas. Hagemeijer, W.J.M. & Blair, M.J. 1997. **The EBCC Atlas of European Breeding Birds: Their Distribution and Abundance.** *T & AD Poyser.* This massive book puts Britain and Ireland's birds into the proper context as those inhabiting some islands on the edge of the Continent – fascinating!

The Historical Atlas. Holloway, S. 1996 **The Historical Atlas of Breeding Birds in Britain and Ireland: 1875-1900.** *T & AD Poyser.* This *tour de force* was produced by Simon using all the historical texts to produce maps for birds a hundred years ago, based on the vice-county system, much used by other naturalists.

BWP. Cramp, S. / Cramp, S. & Simmons, K.E.L. / Cramp, S. & Perrins, C.M. 1977-1994 **Handbook of the Birds of Europe, the Middle East and North Africa (The Birds of the Western Palearctic).** *Oxford University Press.* The nine volumes of the complete work are also available on CD-ROM, there is a concise (2 volume) edition and a journal updating this monumental work. Comprehensive is the word!

NRS, CBC, WBS, CES and BBS reports. These are the ongoing annual investigations of the BTO and are reported each year in one or other of the six issues of **BTO News** published each year. They are the Nest Record Scheme, Common Birds Census (ceasing to be an annual survey with the 2000 season), Waterways Birds Survey, Constant Effort Sites Scheme and Breeding Birds Survey (takes over from CBC).

Birds in the Wider Countryside. Crick, H.Q.P. *et al.* 1998 **Breeding Birds in the Wider Countryside: their conservation status (1972-1996).** *BTO.* This reports on the BTO's Integrated Population Monitoring programme, largely funded by the Joint Nature Conservation Committee. It is expected to be wholly revised every three years or so when the BTO Alert status of each species will be reconsidered.

Birds of Conservation Concern. The NGO Conservation Bodies. 1997 **Birds of Conservation Concern in the United Kingdom, Channel Islands and Isle of Man.** This was put together by the RSPB, Bird Life International, Wildfowl and Wetlands Trust, Game Conservancy Trust, BTO, Hawk and Owl Trust, the Wildlife Trusts and the National Trust. It was designed to act as an *aide memoir* for Government.

UK Biodiversity Action Plans. Two tranches of plans and several reports have been published so far (1999) on behalf of the UK Biodiversity Group by English Nature.

The British List. **The official list of birds of Great Britain with lists for Northern Ireland and the Isle of Man.** This is issued, free, by the British Ornithologists' Union annually – the one used in this book was issued in autumn 1999.

RBBP Reports, Ogilvie, M.A. & Rare Breeding Birds Panel. 1999 **Rare Breeding Birds in the United Kingdom in 1997.** *British Birds:* **92**, 389-428. This is the 25th annual report of the panel and all previous reports have appeared in British Birds. The RBBP has also started to report on non-native species – two annual reports so far and references given in each relevant species account.

Population Trends Guide. Marchant, J.H., Hudson, R., Carter, S.P. & Whittington, P. 1990 **Population Trends in British Breeding Birds.** *British Trust for Ornithology.* A comprehensive synthesis of the Common Birds Census data available to 1988 with commentaries and graphs for each species – a classic.

Irish Bird Report and Book. Hutchinson, C.D. 1989 **Birds in Ireland.** *T & AD Poyser.* This is the book and the annual reports are in *Irish Birds* published by BirdWatch Ireland (the Irish Wildbird Conservancy).Also, in 1999, Gordon D'Arcy published the excellent *Ireland's Lost Birds* (Four Courts Press, Dublin).

Scottish Bird Report and Book. Thom, V.M. 1986 **Birds in Scotland.** *T & AD Poyser.* There is a separate *Scottish Bird Report* as well as the journal *Scottish Birds*, both are published by the Scottish Ornithologists' Club.

Welsh Bird Report and Book. Lovegrove, R., Williams, G. & Williams, I. 1994 **Birds in Wales.** *T & AD Poyser.* The annual Welsh Bird Report has been a part of *Welsh Birds* for several years. This is published by the Welsh Ornithological Society (Cymdeithas Adaryddol Cymru).

Useful addresses of national membership-based organisations you can join

British Trust for Ornithology, The Nunnery, Thetford, Norfolk IP24 2PU. 01842 750050 general@bto.org

Irish Wildbird Conservancy (BirdWatch Ireland), Ruttledge House, 8 Longford Place, Monkstown, Co. Dublin, Ireland. (01 or 003531 from UK) 280 4322 *bird@indigo.ie*

Scottish Ornithologists' Club, 21 Regents Terrace, Edinburgh EH7 5BT 0131 556 6042 *mail@the-soc.org.uk*

Welsh Ornithological Society, c/o Paul Kenyon, 196 Chester Road, Hartford, Northwich, CW8 1LG 01606 77960

The Royal Society for the Protection of Birds, The Lodge, Sandy, Beds SG19 2DL 01767 680551 bird@rspb.org.uk

Index

Entries for species only: bold refers to main entry, and italic to an illustration.